MW00464071

THE BOOK OF STORMS

Other books by Eric Sloane available as Dover reprints

(Log on to www.doverpublications.com for more information.)

Eric Sloane's

Book of
Storms
HURRICANES, TWISTERS AND SQUALLS

Dover Publications, Inc.
Mineola, New York

Contents

A NEWSPAPER publisher once told me that of all daily news, weather tops the list for consistent importance. "Why," he said, "don't you get up a weather feature for newspaper syndication? It should sell at once." That was years ago.

A pile of submitted sketches and polite rejections have taught me an interesting point. "The American man," reads one rejection letter, "is interested in weather all right, but he is a practical person; there is just one thing he wants to know and that is: Will it rain tomorrow or will it not? Weather lore or philosophy? No—that's for nature lovers. The average businessman hasn't got the time for that sort of thing."

The businessman goes his way despite the weather, more so each day. Instead of adapting himself to the weather, his goal is to ignore it. Even aviation is flying around it, above it, or straight through the middle with such speed that weather becomes of less concern every year. If you want to attract a crowd on a busy street corner, just stand there and look at the sky. So few of us look aloft at all, that within a few minutes a crowd will have gathered, staring with you. The upward glance is a scarce thing these days.

So, if a story of weather is to be told to the average person, I thought, and the only thing he "has time for" is its stormy aspects, why not let him have just that? Here, then, is a book of storms, with not a sunny sky to be seen, and with the rumble of thunder every time you turn the page. But if it gets you to look upward more often, it will have accomplished a purpose. For then, when you are weatherwise and aware of the sky, perhaps you will be more prepared for a philosophy of good weather. And that is what I want my next book to be about.

ERIC SLOANE

THE BOOK OF STORMS

AT 1,000 feet, man begins to disappear. At 3,000 feet, where the lower clouds begin to form, he has already disappeared. At 10,000 feet, highways turn to threads and the ants that were automobiles, slacken their crawling and vanish quickly into tininess. At 20,000 feet, where storms are born, whole villages are swallowed by the mosaic of earth, and man becomes, for the moment, only a memory.

From aloft, the immensity of the atmospheric ocean makes human activity seem insignificant. Weather, which we had regarded as being created for our benefit, becomes part of global machinery and man merely the tenant. We atmospheric creatures are so unaware of the sea in which we live that it takes a storm to discipline us into the acknowledgment of its existence.

A most unfortunate human trait is the ability to forget

The Effects of Storm

the good and remember the bad. It is like that with our regard toward the weather. You will seldom recall the good days, yet the most bygone storms will leave an indelible impression on your memory. The wonders of good weather, when reduced to meteorological-book explanation, seem mathematical and dull; yet, any story that includes a weather scare is instantly exciting. Perhaps it is because we so little understand storms that we stand in such awe of them. What a pity to forget that weather is more often good than it is bad!

Fair-weather accidents are, naturally, never blamed on the weather. Yet when newspapers publish a list of storm fatalities, it includes anyone who slipped on the ice, had a heart attack while shoveling snow, or drowned while risking his life during a flood. The word weather itself has come to mean unpleasantness.

This blame-the-weather attitude makes the average person think that chaos and destruction are the only fare that storms have to offer us. It causes little children to cringe and hide from what they might possibly enjoy. A bolt of lightning may make a child storm-shy for the rest of his life, but lightning has killed far fewer people than have fireworks! A thunderstorm is more exciting and beautiful to behold than a Fourth of July display, and not entirely a menace. Storms, like rainbows and sunsets, are special performances in the theater of weather.

Those strange and wonderful people of England derive enjoyment even from rain and fog. The English have thunderstorm census-takers who count the number of lightning strokes, and citizens write letters to newspapers to tell of their personal experiences and observations during a storm. They have learned the secret of enjoying bad weather as well as good.

The best way to lose fear of storms, it seems, is to get better acquainted with them. The machinery of weather is not supernatural but as logical as the mechanics of your automobile. It is nature in its unpredictable ways that steers the weather machine along its erratic routes. As unpredictable and inconveniencing as weather can be, we might recall that life would be dull without some uncertainty. Life on earth, by the way, would cease entirely if there were no weather changes at all.

Before a storm, animals and insects become nervous, often irritated, sometimes even vicious. Flies bite, fishes become bold, birds dress their feathers and quarrel with one another; horses get quick-tempered and are most likely to bolt. Lowering storm pressure releases gases and odors that stimulate animal sensitivity; dogs become nervous and alert. It is natural, then, to presume that humans might also react to prestorm weather. Quite so: Rising humidity and lowering pressure affect us both physically and mentally, but so mildly that only slight nervous reactions are produced. The general effect of prestorm air is to cause restlessness in all human beings. Very interesting and most important, however, is the fact that this restlessness may be either good or bad, according to the individual. Extreme lowering of pressure is exactly like alcoholism in its effects. It might make one person pugnacious and it might make another overly carefree. It might make us amusingly happy, or it might cause us to be unnecessarily depressed, according to the state of our immediate physical being plus, of course, our personality. The only difference between very low-pressure effects and whisky is the absence of a hang-over.

Most creative people find that their work is benefited by storms. You will often hear an artist or a writer say that he can do his best "near a fireplace, set snug inside while the weather is roaring outside." Their ideas are stimulated by the very slight lessening of oxygen in their arteries, possibly to the extent of one glass of wine. A drink of liquor before or during a very low-pressure storm, incidentally, would have more effect than it would in good weather.

Later in this book, you will see how the approach of a

storm changes the electrical charge of the land below it. This phenomenon should not exempt human beings. Although our electrical potential is not a known factor in its relation to our physical being, it does seem to affect our mental activity and our nervous system. You will notice that walking across carpets on a cold dry day will accumulate enough electrical charge to "spark" anything you touch. A moist day, you must agree, should do the opposite. Does an overcharge of electricity stimulate energy? At least we know that a dry day makes us energetic while a moist day tends to make us sluggish.

The approach of a storm creates many changes in the human being which he cannot explain. He searches for an outside change, while the change is within himself. Things "sound different" before a storm. They even seem to "look different." Few of us will deny that the sound of train whistles or the drone of an airplane is different just before a rain. Others can sense a coming storm by a sensation of ache in rheumatic joints or old scars. To date, this is mostly theory and although every doctor will acknowledge it, and some even write papers on the subject, the actual proof is yet to come. It is logic that a difference in the moisture content, oxygen content, and electrical potential of the human body, even the slight difference that an approaching storm might cause, will cause some appreciable effect.

A thinning of air will change sound because sound is no more than atmospheric vibration: thinning air will cause thinning vibration. As you ascend to low-pressure altitudes, sound diminishes and, at stratospheric heights, sound has a mushy "echo effect." On the moon, where there is no atmosphere, you could shoot a gun next to your ear, yet hear nothing. People who can "hear rain coming" are not speaking unscientifically.

Lowering pressure releases odors that were hitherto captive under good-weather high pressure. That, and an in-

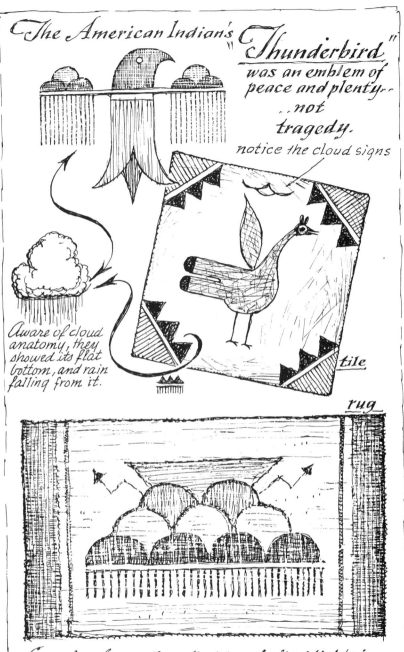

The American Indian's "*Thunderbird*" was an emblem of peace and plenty — ..not **tragedy**.

notice the cloud signs

Aware of cloud anatomy, they showed its flat bottom, and rain falling from it.

tile

rug

Cumulus formation, flat top, forked lightning: the ancient Navajos knew their Storm Clouds.

crease of moisture, will make things smell richer and clearer just before a rain; so those who "smell rain" are, again, telling the truth.

Children sense the tension of grownups during a storm and learn a fear of wind and rain at an early age. Indian children, who were taught that thunder and lightning were works of a kind and just God, actually looked forward to seeing the heavenly fireworks. The thunderbird emblem of the southwestern American Indian is far from being an evil or frightening sign: it represents peace, plenty, and happiness. Children should learn that storms are devised for other reasons than bad, that storms are neither accidents of nature nor attacks upon life. Drownings and other accidents that occur during storms are caused more often by fright and panic than by wind or rain.

Wind is the great pruner of the forest. It removes weak trees and diseased limbs, so that only the strongest trees may have room to grow. Trees assume their shape and bearing by exercising, just as any other living thing. A child that visions a forest doing its "exercising" during a big blow might be less frightened or at least more amused than one who remembers such lore as "when the bough breaks, the cradle will fall," etc. A child that is taught what makes the thunder and lightning and wind will also become a less nervous person. It seems silly, but there are still adults who hide scissors and turn mirrors down so they "won't attract lightning." Others believe that thunder curdles milk, or that stepping on an ant will make it rain, and so on into the maze of ridiculous folklore.

Sudden storms have dramatic and unforgettable entrances. Like the performances of great actors, they seem to be staged just for your very benefit. Every near-miss bolt of lightning seems to have been aimed right at you. However, in parts of Java, where thunder is heard on an average of over three hundred days a year, there is very little dread or nervous tension connected with storms. Dr. C. E. P. Brooks,

8

who once collected data in Buitenzorg, Java, found it to be the most thundery place on earth. He estimated that, at any given moment, close to 2,000 thunderstorms are in progress somewhere upon the earth, with about 40,000 of them occurring daily.

Storms are watched a bit closer nowadays, since the controversy about drastic changes in our weather. Actually, there seems, however, to be as much reason for remark upon the fact that the change has been moderate during the last century. Considering the melting of glaciers and the existence of weather cycles, the weather could have changed even more. The average temperature in the heart of troubled New England has risen only about three degrees in the past hundred years.

There are those who say that the depletion of forests, urbanization, and natural heat of civilization are partly responsible for a rise in temperature; but the meteorologist gives no credit to such a theory. The weatherman, who scoffs at the idea, still includes, in his temperature reports, such remarks as "slightly cooler in the suburbs," and "colder in the outskirts." It does seem that a thermometer set up in the small wilderness-surrounded town of a hundred years ago might show a three-degree lower temperature than the same thermometer located within the twenty-mile diameter of the metropolis of today.

Anyway, let's look at that change which has been a topic of so much discussion by observing the monthly mean temperature of New Haven (taken by Elia Loomis of Yale College) in 1867 and comparing it with that of eighty-five years later.

The chart does not look too drastic. Ever since it was compiled, winters have been "picking up," and indications are those of a cycle ending. The worry about palm trees flourishing in New England, and coastline cities being flooded by the sea within a thousand years is pure weather hysteria.

	Then	Now	Rise
January	26	29	3
February	28	29	1
March	36	37	1
April	46	47	1
May	57	58	1
June	67	69	2
July	71	72	1
August	70	70	0
September	62	64	2
October	51	54	3
November	40	43	3
December	30	32	2

The effects of storm upon man's living and thinking will always be a controversial subject. A stormless place might seem to be utopia, yet a planet where there could be no storms would have to be a planet where there would be no air at all—and, therefore, a planet with no life. The final observation comes down to the difference between a moral and a material view of the effects of storm. Like war, which is disastrous on one hand, yet productive of stimulated endeavor, weather both destroys and builds up. Lightning fertilizes the earth to the extent of about a hundred million tons of nitrogen a year, yet it is also the cause of over half our forest fires.

Switzerland was held up to an old-school Army general as being a warless, peaceable, and, therefore, progressive nation. "Yes," retorted the general, "but the only thing they have to show for it, is the cuckoo-clock." Not a very true bit of humor, of course, but it demonstrates the point. The climatic utopias of the world are so often the least inhabited and with the least architectural culture to show for it: the areas of brisk, even merciless weather changes are teeming with inhabitants who have built accordingly, to combat the elements.

It is not at all farfetched to compare weather with human life, for few things in our universe are so identical. We are born mysteriously into the world, very much like clouds, and we disappear back into the world just as clouds disintegrate into the atmosphere from which they came. The sky is as changing as human passions, and as spiritual in its ways as our own emotions. Mood music composed to give the impression of a storm, could as well tell of a comparative human mood. Shelley was describing a cloud by comparing it to human life when he wrote these meteorologically correct lines:

I am the daughter of earth and water,
    And the nursling of the sky;
I pass through the pores of the ocean and shores;
    I change, but I cannot die.
For after the rain when with never a stain,
    The pavilion of heaven is bare,
And the winds and sunbeams with their convex gleams,
    Build up the blue dome of air,
I silently laugh at my own cenotaph,
    And out of the caverns of rain,
Like a child from the womb, like a ghost from the tomb,
    I arise and unbuild it again.

The sky, which is backdrop for the theater of our everyday life, is fair most of the time. In being stormy now and then, it is not too remote a reflection of the play.

Storm Mechanisms

It is certainly strange how the course of events evolves. Many years ago, while rowing on a lake with a little boy, we had to pull into the shelter of a boathouse during a shower, and there we talked about weather. I told him how water and air were so much alike, how both things reacted the same, and how scientists use water to simulate experiments with atmosphere. I also told him how there were three major ways in which a storm may originate.

The little boy was like all little boys and very logical; he handed me one of the oars and said, "Go ahead and show me—make a little storm in the water for me."

For a second I was stumped, but you can't let a little boy hang in mid-air like that, so I took the oar and tried to show him, using it to stir the water and demonstrate storm mechanics. Little did I dream that years later I would use that illustration in a book. But with apologies to the more learned men of meteorology and with hopes that they will forgive me, I use the illustration as a crude but simple way again to put over the point.

There are many variations of each mechanism, but the three kinds of disturbance that you can "kick up," either in water or within the atmosphere, are: *stirring, pushing,* and *lifting.* "Stirred" storms are called *cyclonic,* after a Greek word meaning "coiled." "Pushed" storms are called *frontal,* because the front of the pushing air is where the collision and rain occur. "Lifted" storms are called *air-mass, thermal,* or *orographical,* according to what caused the sudden lifting of atmosphere.

We will take each of these mechanisms, one at a time, and see what kinds of storm they create.

Cyclonic Storms

THE first of a thousand and one confusing things about weather is: why wind from a storm doesn't come from the direction of that storm. The second is: why high pressure doesn't blow outward and directly into low pressure (like the air from a balloon does), into the surrounding atmosphere. The explanation is that we are used to measuring with earthly "yardsticks," whereas the laws of space make earthly things look illogical. To us, the horizon is flat, but from the moon, I guess that same horizon looks pretty round.

To see one illustration of the difference between earth conception and space conception, look at the opposite drawing. It shows you at the North Pole, directing a missile toward the equator. The missile goes true, according to Newton's laws of motion. But without your realizing it, both you and the target have moved! You have turned, and the target has moved in a circle around you, with the turning earth. If you could see the missile's flight, even take a motion picture of it, you would still see a perfect arc, veering to the right. According to space mathematics, or to the man on the moon, the bullet made only a straight flight; but to every person on earth, the arc is so real that when we shoot a long-range gun, we have to aim from ten to fifty feet to the right of our target.

All this effect is reversed in the southern hemisphere. By looking at the drawing and changing your position to the South Pole (thereby reversing the earth's turn), the arc would reverse also. The bullet would then veer to the left! And storms, you will learn, turn one way in the northern hemisphere, the opposite way in the southern.

When wind goes from high pressure to low, it gets involved with global mathematics; although it gets from high pressure to low, what it does more than anything else is just whirl around its destination. It whirls primarily be-

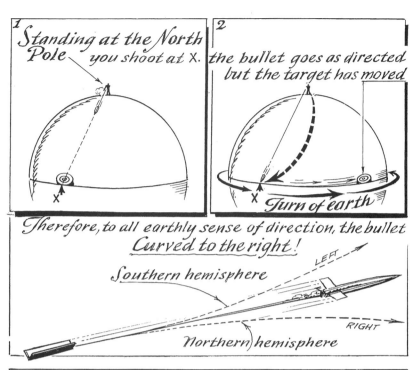

1 *Standing at the North Pole* — you shoot at X. the bullet goes as directed

2 but the target has moved

Turn of earth

Therefore, to all earthly sense of direction, the bullet <u>Curved to the right!</u>

Southern hemisphere

LEFT

RIGHT

Northern) hemisphere

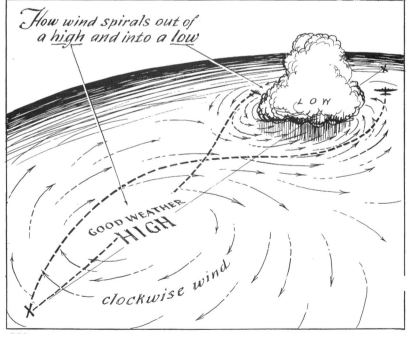

How wind spirals out of a <u>high</u> *and into a* <u>low</u>

LOW

GOOD WEATHER HIGH

clockwise wind

X

cause of the turning of the earth, and also because fluids (like air and water) revolve at the drop of a hat. When you let the water out of the tub, does it just pour out? No, it whirlpools out! It will turn whichever way the contours of the basin or the movement of your hands made it start, but whichever way it does start, it keeps right on until the basin is empty.

As soon as any mass starts moving, and it is big enough to divorce itself from its earth contours, global law takes over. It says that northern-hemisphere winds around a low pressure go counterclockwise and winds around a high go clockwise. The exact mathematics for this is not essential in our picture book of storms, but it is important that you know it. The drawing of wind spiraling out of the high pressure into the stormy low demonstrates the pattern for you. The dotted lines show routes past the storm of two airplane pilots. One confronts only head winds while the other gets helpful winds. The smart pilot knew that he should pass such storms on the right to take advantage of their counterclockwise wind.

The turn-to-the-right law is known as the Coriolis force (named after the French mathematician Gaspard Gustave de Coriolis, who discovered this law). Although it affects everything on earth, it is only evident in large-scale masses or long-range movements such as global masses of air or large bodies of water. The hurricane, for instance, is so big that it is almost completely controlled by Coriolis.

The tornado is fairly large up where it forms between two air masses, but it is often too small to be controlled by Coriolis force; so out of every thousand tornadoes which are all supposed to turn counterclockwise around their low pressure, about six will turn clockwise. Oceans are affected by earth motion, all having currents going clockwise in the north and counterclockwise in the south. The North Atlantic makes a complete circle in about three years.

The word cyclone is not another name for hurricane or

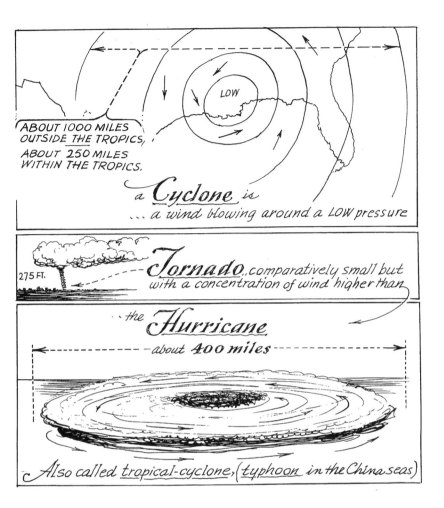

ABOUT 1000 MILES OUTSIDE THE TROPICS, ABOUT 250 MILES WITHIN THE TROPICS.

LOW

a *Cyclone* is
... a wind blowing around a LOW pressure

275 FT.

Tornado, comparatively small but with a concentration of wind higher than

the *Hurricane*
about **400 miles**

Also called tropical-cyclone, (typhoon in the China seas)

for tornado. It was originally a meteorological word that described the motion or direction of wind; it had nothing to do with velocity. In fact, meteorologically speaking, a cyclone can be very slow, even zephyrlike. There doesn't even have to be rain. A cyclone in strict weather language is simply a wind revolving around any low-pressure area. But the strongest winds have always been associated with low-pressure storms, so the word cyclone has come to mean just what most people think anyway. Instead of tornado cellars (which is what they are), we still call them cyclone cellars.

FACING THE WIND, THE NEAREST STORM IS USUALLY TO YOUR RIGHT

because

STORM-WINDS PINWHEEL COUNTERCLOCKWISE AROUND "LOWS."

LOW

When you point, aim slightly to the Rear.

The old rule about facing the wind with your right arm out, in order to locate the nearest storm, has seldom failed. It is a good way for checking up on the movement of a hurricane, and also for analyzing the weather map. Actually, the pattern of wind is also toward the low's center; so in pointing outward, do not point directly to the right but slightly backward too. If the wind is too calm to feel, you can always refer to the clouds and the direction in which they are traveling.

THE tornado is a frightening piece of weather machinery. Add to it the sound of a thousand freight cars rolling by and the continual thunder of cannon. It is the world's smallest and most violent storm. Its path ranges from about two hundred feet to a half mile wide; our illustration shows only average dimensions. It has been estimated that the wind within the vortex reaches about five hundred miles an hour, even eight hundred.

Tornadoes generally occur in the southwestern portion of a cyclonic area, which possibly is the reason they usually travel from that direction. They usually exist from only a few minutes to about an hour, and the snakelike appendage which is solid cloudform condensed around the vortex of the whirlpool of wind is crackling continuously with lightning. Called a tornado funnel, it dangles like a breathing snake, bulges large first at the top and then at the bottom, writhes into almost any shape, then whips back to its anatomy of destruction as seen in the drawing. Before it came in contact with the earth, it could have been mistaken for a large rope dangling down from the parent cloud.

Few people have gotten close enough to learn what the inside of a tornado looks like, but from those few, we understand that their funnels were about seventy-five feet in diameter. They were all lit inside by lacelike lightning, and exuding a screaming hiss. The noise around a tornado is so deafening that there is always some argument as to whether there was thunder within its funnel. It is presumed that the constant lacework of lightning causes more of a continuous roar than the explosions of ordinary thunder and that the tornado's "roar of wind" is really a continual noise of electrical discharge.

Anatomy of the average Tornado.

Lightning

funnel 2000 ft.

about 275 FT.

N

W

E

S

Moving to N.E.
20 to 50 m.p.h.

An ancient name for tornado was whirlwind. Isaiah mentions "... and the wind shall carry them away, and the whirlwind shall scatter them" (41:16). But the word tornado, which is supposed to come from the Latin "to turn," seems also to come from the Spanish word "tronado," meaning thunderstorm. There is no doubt about the mechanics of the tornado having been known since Biblical times, and no other phenomenon of weather has been as carefully recorded.

In June, 1822, near the mouth of the Ganges, 50,000 people lost their lives in one tornado. Forty years before, one passed over the Barbados and it took 4,000 lives. In those days, of course, there was little record of tornado activity in the American Middle West because it was then uninhabited except by Indians. But there are good accounts of tornadoes in New England. The drawing illustrates an account of one which passed over New Haven in 1839. "It advanced in a direction N. 50° East. On the right side of its path, the prostrate trees were uniformly inclined toward the north, while on the left-hand side many were inclined toward the south." The primitive drawing with "fallen trees" indicates that the tornado dipped for an instant, sucked the surrounding trees down, and then lifted itself away.

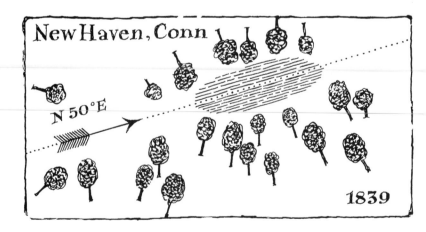

New Haven, Conn

N 50° E

1839

Tornadoes have been reported in every time of the year, and in every state. In the South, they are most frequent during very late winter. The Middle West gets its tornado season during the spring, when tornadoes are watched for in May and June. Kansas, alone, has probably had close to a thousand tornadoes to date and it rates as the number one tornado state.

Recording low pressure within a tornado is difficult, but it probably lowers to about twenty inches. This measurement is unimpressive in round barometic figures, but it means that a house with normal good-weather pressure inside would suddenly have an outward force of many tons upon an average-sized window, possibly a few thousand tons within the average house. Of course, it would explode outward. A human being directly in the tornado's path suddenly blows up like a balloon and ruptures to death. Automobile tires burst. Barns push their complete walls outward and lay them down neatly on all four sides, often leaving the livestock inside unharmed. Wells are sucked dry of water. Canned goods explode and sealed bottles burst. Airtight watches shatter their crystals.

People of the Middle West tornado belt seldom try to side-step a tornado because they usually have a cyclone cellar handy. Some of the first Middle West schools were built as cellars to begin with, both for protection from tornadoes and for coolness during warm weather. One such underground schoolhouse near Lawrence, Kansas, saved its entire group of children by having been below ground; the only things blown away were the outhouses which were above. One of the first things Kansas farm children used to learn was to "run southeast or northwest" if caught on foot in a tornado. This would get you fastest out of the way of a northeasterly moving tornado.

It has been recorded that tornadoes occur within hurricanes, but the evidence is always difficult to ascertain; a hurricane is big enough and rainy enough to blot out most

This is our Tornado Playground

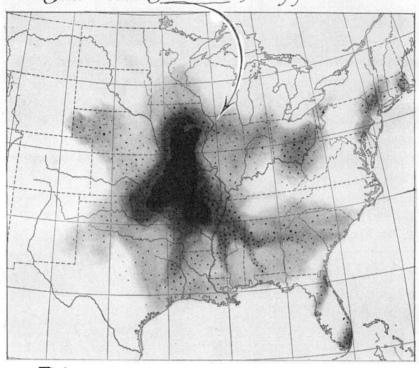

These are the times to expect them by months.
Average frequencies per month.

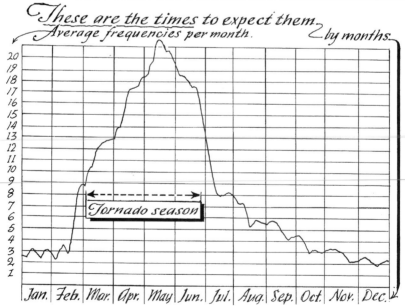

Tornado season

of its internal machinery from view. Tornadoes have occurred on the fringe area of hurricanes, however, where they can be definitely seen. Using water again as our medium, the drawing recalls that whirlpools often do occur along the rim of large swirling river masses. Perhaps the comparison is scientifically vague, but the picture is most demonstrative to the layman of how several tornadoes might form along the fringe of a hurricane.

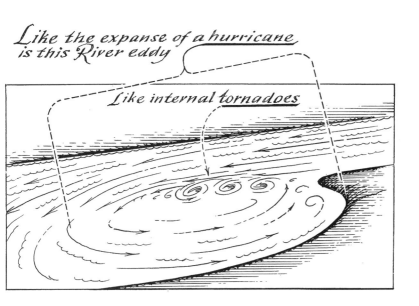

Like the expanse of a hurricane, is this River eddy

Like internal tornadoes

Tornadoes most often happen in conjunction with cold-front thunderstorms. The next time a fast-moving summer thunderstorm approaches, do look up into the swirling line-squall clouds that blacken the sky just before the rain starts. You will usually see dark lumps of cloud hanging down, lit around their edges and blackest in the middle. It is from such mammato cloud formations that the tornado funnel emerges.

When warm, wet Spring air tries to "mushroom" through an upper stream of cool air—the tornado-trigger (4) is born.

upper cold air

warm, wet air

dry, cold air

It takes the warm wetness of spring and the cold dryness of persistent winter air (or the equivalent of such extremes) to spring the tornado trigger. It could be sprung by any number of conditions, even by the atmospheric disturbance above a burning forest or an oil field. But when an upper cold front sweeps into a thunderstorm area, the odds go up immediately. By watching thunderstorm areas and matching them up with upper cold air masses, weathermen have been able to predict tornadoes for the first time. Exactly where they will form and what path they will take is impossible to predict, but such seasonal tornado areas can be announced as much as a day or two ahead of time.

Tornadoes are so intense, yet so small in size, and last for such a comparatively short time, that research into them is difficult. Even their velocity is too high and quick to record.

Recently, however, a tornado described a circular coil of lines on the ground exactly like old-time penmanship exercises. By measuring the coils and multiplying them by the speed of the tornado path, a velocity of four hundred miles an hour was shown.

  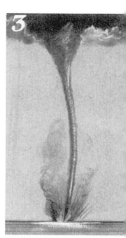

Water-spouts

A WATERSPOUT is actually a tornado above water. Tornadoes are fed by heat and moisture; the right amount of both makes them machines of great destruction. Whereas dust devils and prairie whirlwinds have not enough moisture to grow to full tornado fury, the waterspout has too much water at hand and not enough heat to make it a first-rate killer. True, some waterspouts have tossed things around, but the average one does little more than tear the flags and sails of a boat that might go through it.

Waterspouts have occurred in rivers and lakes, sometimes starting on land and floating out over the water, carrying dust and leaves of trees with them. The cross-section drawing shows that although many people think the stormspout funnel consists of sea water sucked upward, it is really composed of thick fresh-water cloudform. Sea water is seldom carried up more than ten feet and the foam-shelf sucked up around the seaspout funnel is seldom over two feet high. The cylinder of spray around the funnel consists of smoke-like condensation within the outer diameter of the whirlwind. When the seaspout disappears, the funnel goes directly back into the atmosphere and does not "fall as a shower of sea water," the way most people believe.

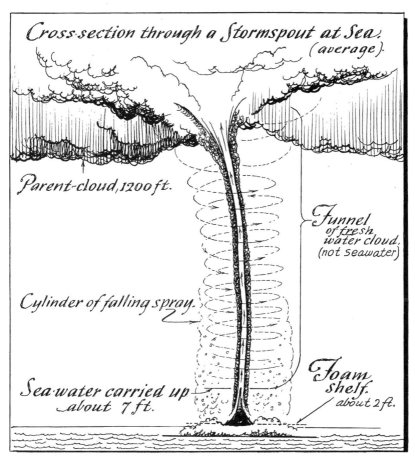

Cross-section through a Stormspout at Sea.
(average).

Parent-cloud, 1200 ft.

Cylinder of falling spray.

Sea-water carried up about 7 ft.

Funnel of fresh water cloud. (not seawater)

Foam shelf. about 2 ft.

  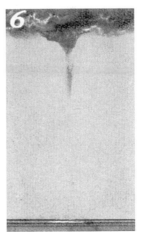

Prairie whirlwind, and dust-devils

Desert Sand-pillar 500' to 1000' high

50 to 500

Whirl-winds

WHIRLS may be sent up within the atmosphere wherever heat causes intense updrafts. Whereas a tornado is born at the cloud level and grows downward, the dust devil or prairie whirlwind grows upward from the ground. There is more heat than moisture in this mechanism, so the proper ingredients for dangerous tornado velocities are again not there. On the hot prairies or deserts you might see several whirls going at one time; in summer they are almost so common that they are part of the landscape. Only a few reach full size.

Every city street has its miniature dust devils where moving winds and rising heat combine to whirl newspapers and dust upward. They are all too small to be ruled by Coriolis force, and they turn either to the right or the left.

On the desert, where whirlwinds reach maturity, they are known as "pillars of sand," which sometimes last for hours and reach mountainous heights. When they occur above an extensive fire, they are called smoke devils or fire whirlwinds. An account described such a whirlwind above the conflagration of Moscow in 1812. The flames were sucked into the vortex and it stood for an hour "like a flaming hourglass, a half-mile in the sky."

IN ARID PLACES, where the surface earth has been reduced to dry dust, a sudden mass of rising warm air sometimes triggers a storm-size whirlwind of disastrous consequence. The western states have had updrafts carrying ground dust over two miles high, making a solid wall of darkness. Dust showers have fallen in Vermont and New Hampshire that originated in Texas, Oklahoma, and Kansas, about 1,500 miles away. It was estimated that a ton of lime was deposited on every two square miles where such dust settled, which was probably one of the longest distances involved in natural ground fertilizing.

Dust Storms

The Italian Riviera once had a "red rain" which sent people into a panic of superstitious hysteria. It proved, however to be caused by a previous storm of red dust from the African desert, which became nuclei for the rain, and dyed each drop a bright red!

The ability of wind to raise dust is understandable when you look at the drawing. When you blow along the top of a sheet of thin paper, it lifts upward against the moving air instead of blowing downward. The low pressure caused by moving air, lifts dust vertically just as the air on top of an airplane wing lifts the wing upward. The harder the wind blows, the more lift there is, and the more dust rises.

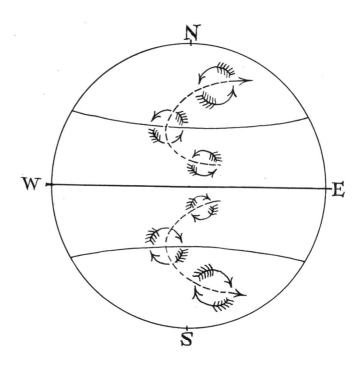

Hurricanes   **T**HE drawing above is a hundred years old. It is from a textbook that proves the old-timers were almost as familiar with hurricane theory as we are today. The pattern of rotation is correct, and the curved routes of the rotating storms could be a general average of the ones shown in the opposite drawing.

With a hundred-mile core of thunderclouds and torrential rain, the hurricane is like a squashed doughnut that spreads out as far as three hundred miles from its hole in the middle. It is designed like a cyclone such as the mild ones we experience in the daily pattern of extra-tropical weather in the United States, but instead of touring about gently, it goes into high gear, opens the throttle all the way, and throws away the key. It is a pinwheel of Coriolis force bigger than any other weather in its way. Nothing could stop it except another hurricane. Sometimes two hurricanes

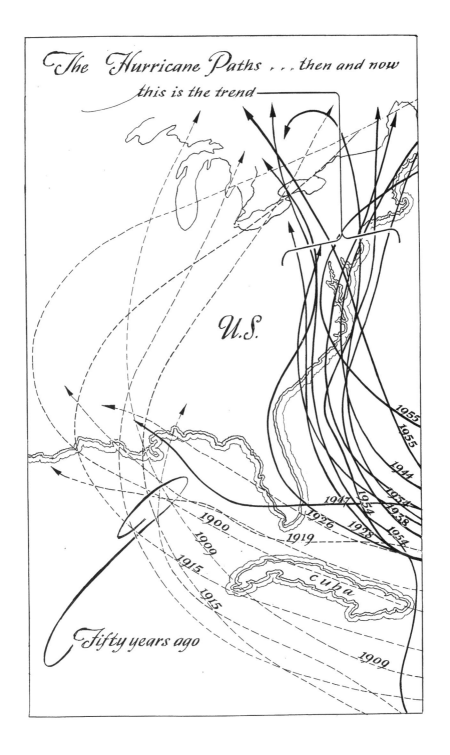

The Hurricane Paths . . . then and now

this is the trend —

U.S.

1955
1955
1944
1954
1954
1947
1938
1926
1954
1928
1900
1919
1909
1915
1915

Cuba

Fifty years ago

1909

Looking down at a Hurricane

Swells radiate outward from the storm center.

do collide and they merely bounce off at different angles, with only a little of their force diminished.

Why hurricanes are born is still a mystery. They are children of the hot, calm air near the equator. If this rising wet heat occurred directly upon the equator, Coriolis would not have been concerned and you would simply have some warm air ascending and other air rushing in to take its place. But the fact that this occurs near the intense rising heat of equatorial calm, yet just far enough away from it for the spinning force to enter the picture, is what starts the hurricane machinery in motion. It might take a few days or a week to gain momentum and to grow in size. Until then, it usually stays in its own back yard. But as soon as it becomes a grown hurricane, it starts to leave home. Its favorite routes are shown in the drawing. The dotted lines show how hurricanes wandered during the first twenty-five years of this century: the solid lines show the hurricane routes of the next twenty-five years, sweeping abruptly northward toward New England. Only the next fifty years will prove whether this trend is a cycle or a permanent change.

If we were to count the written records of people killed by West Indian hurricanes since the time of Columbus, the list would come close to a million, but no one will ever know the true total. The nineteenth century brought more accurate records, and the hurricane our map starts with (in the year 1900) killed 6,000 people and did 20 million dollars' worth of damage. Since the 1928 hurricane, which killed close to 2,000 people in Florida alone, the death lists have been getting lower because of radio warning systems.

The cloud warning of a hurricane is very similar to that of a warm-front storm, inasmuch as its high layers of clouds begin weblike and thicken to rain clouds as the ceiling lowers. The barometer lowers too, and the drawing shows why. Air pressures and barometers are strange to many people who know nothing about them except that "when

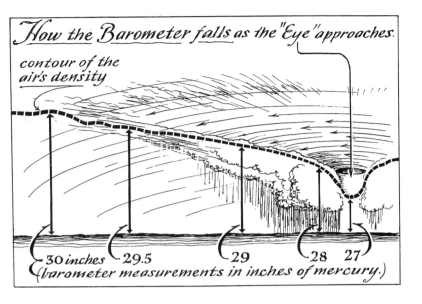

How the Barometer falls as the "Eye" approaches.

contour of the air's density

30 inches   29.5   29   28   27
(barometer measurements in inches of mercury.)

the barometer falls, a storm is near." It might simplify things for those people, to think of air pressure as the weight of air; and to think of the barometer as its weighing machine, or scales. The analogy is not too far from the truth. When a mountain of air density passes overhead, the "scales" will weigh it by using a higher amount of balance (mercury). When a valley of air density passes overhead the scales need less balance to weigh it, and a lower height of mercury is therefore used. As the center of the hurricane's atmospheric sinkhole approaches, the barometer begins to record less and less density overhead. As the drop quickens, the rush of wind around and into the "hole" quickens too.

A plain barometer is like a clock and it leaves no record behind; you just have to keep watching it. A barograph, however, inks a continuous line of pressure and it presents you with a graph, like the dotted line in the drawing. Oddly enough, the barograph line outlines the anatomy of a hurricane, just as an artist might draw it.

Cross section through part of a Hurricane.

winds in knots

50 55 60 70 80 90 100 12

very gentle
downdraft

130   135   130   0   0   130   135

Intense
updraft

Eye

Intense
updraft

By watching the barometer for its lowest drop, and using the method of facing the wind with your right hand pointing outward, you can pinpoint the hurricane's center, learning if it is past you or still on its way. If you are directly in the path of the eye, you will then get the lowest barometer readings. On the fringe of the storm, the readings will be less severe. This is shown in the cutaway drawing where a hurricane has been sliced into three barometer readings from three locations. Below, you see an average barograph reading from a location in the center of the storm path as in case (B). The dotted line shows how wind increases toward the center, suddenly changes to a zero within the eye, only to start again in a reverse direction on the opposite side of the calm.

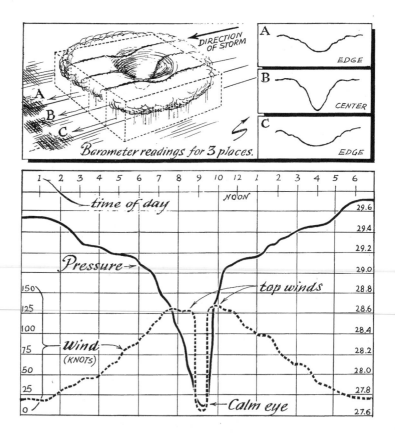

THE strangest experience of any storm is probably to behold the eye of a great hurricane.

Just as you are becoming numbed to the fury of wind and rain, an untimely lull will occur. The roar of storm recedes. The noise of rain, like continuous thunder, moves away like an echo. The unnatural quiet which takes over makes the hurricane of a moment ago seem a memory. The rain stops now, but an uncertain drizzle continues.

A sudden change of temperature dries the air, and you have the sensation of stepping into an oven. It is like being in the heat of the midday sun without there being any sun. There is no wind. At once, the sky clears and you catch glimpses through distant cloud fragments of your position within the storm. You are within a great quiet amphitheater of dark cloud walls. You are at the bottom of the well that is the hub of the hurricane. Millions of square miles of storm are now moving around you. The world's biggest machine of destruction is centered around you, yet it is quiet where you stand!

But not for long. The average hurricane eye is fourteen miles across; this one was probably about half that size. The distant moaning of wind comes into being again, and one side of the amphitheater looms close. Almost at once it overtakes you with boiling clouds, and like stepping out of a shelter, into the storm, you have entered the black tempest of the opposite side of the hurricane's eye.

The revolving wind at the other side of the eye, of course, will now come from the opposite direction. Often, trees that were blown down will be lifted up again after going through the eye, and then be blown down in the opposite direction as the eye passes.

In 1912 the Rev. J. J. Williams, S.J., wrote an account of his experience in the hurricane at Black River, Jamaica. "Then succeeded a breathless calm," he wrote, "that seemed to indicate the very vortex of the storm was passing over us. This lull lasted for about three hours. The unnatural stillness, marred only by an occasional drizzle, was itself portentous of approaching trouble. As there had been no change of the wind, the knowing ones prepared for the worst. . . . The rain was coming in fitful gusts, when suddenly we seemed to be standing in the midst of a blazing furnace. Around the entire horizon was a ring of blood-red fire, shading away to a brilliant amber at the zenith. The sky, in fact (it was near the hour of sunset), formed one great fiery dome of reddish light that shone through the descending rain. . . . Then burst forth the hurricane afresh, and for two hours or more (I have lost track of the hours that night), it raged and tore asunder what little had passed unscathed through the previous blow."

Cross-section through The Eye

Slow downdraft

Calm

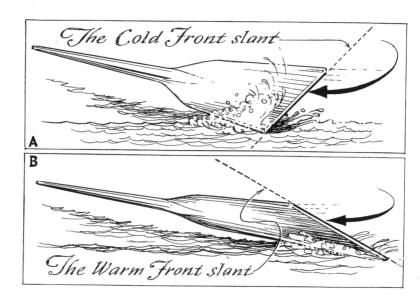

The Cold Front slant

A

B

The Warm Front slant

**Frontal
Storms**

MANY people believe that weather changes will be just present air becoming wetter or drier or warmer or colder. They think that for tomorrow to be warmer it must be to-day with more heat added to it. Weather, however, does not work like that. It is caused by invading masses of air that move over the face of the earth, changing atmospheric conditions wherever they go. Naturally, these air masses are wetter when they come from over water and drier when they come from over dry lands. They are also colder when they come from the north and they are warmer when they come from the south. If you are an air connoisseur, you can actually tell where air came from by just sniffing it!

The air you breathe in as you read this was a few hundred miles away yesterday; the same air will be a few hundred miles in the other direction by tomorrow! Air masses of different air may flow over and around us without discomfort or storm, but their *fronts*, like the front of any attacking object, are what makes the "bang" and causes frontal storms. The fronts of air masses on the march are what

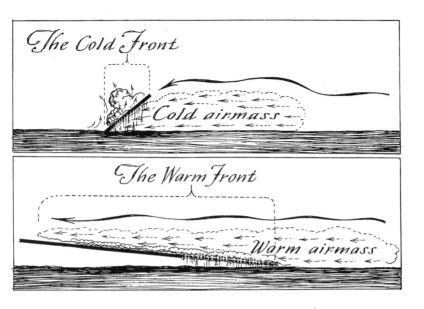

weathermen look for; that is also what weather maps feature.

Going back to our "oar and water demonstration," let's look at frontal or "pushed storms" and see how by pushing air or water you make a disturbance or splash. The front of a fluid mass is never vertical. As the first drawing shows, there are two slanted ways to push water, according to the angle you hold your oar. The first (*A*) is similar to a cold-front angle and second (*B*) is similar to a warm-front angle. The "snowplow slant" of the cold front makes the bigger type of storm disturbance, while the receding warm-front slant is gradual and makes less disturbance. On this page you see actual warm and cold air masses on the march, and the slants of both their fronts.

If you wonder why warm air should make a different frontal angle from cold air, do remember that warm air is light and that cold air is heavy. Light warm air flows up and over the atmosphere in its path, while heavy cold air noses down and roots it into updrafts and cumulus storm clouds. The slant of a cold front is so steep that you can often visualize it, sometimes see it, in an approaching storm. The angle of a warm front, on the other hand, is so slight (often 200 to 1) that an automobile would not roll down such an imperceptible grade: it is so big, often reaching a thousand miles or more, that a person can only see a bit of it at a time and you could not see its complete anatomy ever. The warm front pulls overhead like a slanting ceiling, while the cold front comes at you at a backward angle—like a snowplow. Looking at the drawing which compares the two fronts, notice that the fullest precipitation from the cold air mass falls *behind* the front, while the precipitation from the warm air mass falls *before* the front.

One of the identifying differences between warm and cold air masses and their fronts is in the clouds. Cold air masses

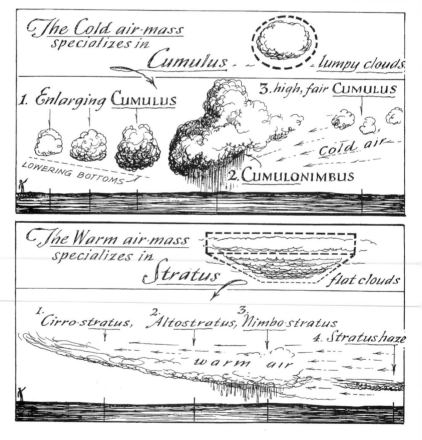

The Cold air·mass specializes in *Cumulus* . lumpy clouds.

1. *Enlarging* Cumulus   3. *high, fair* Cumulus

LOWERING BOTTOMS   2. Cumulonimbus   Cold air

The Warm air·mass specializes in *Stratus* flat clouds

1. *Cirro·stratus,*   2. *Altostratus,*   3. *Nimbo·stratus*   4. *Stratus haze*

warm air

usually specialize in cumulus or lumpy clouds, while warm air masses specialize in stratus or flat ceilinglike clouds. The cold air-mass storm approaches in fields of growing *cumulus* clouds; it becomes a great mass of *cumulo-nimbus* mountains, and when the storm is over it leaves high, very white fair-weather *cumulus* behind it. The warm air-mass storm approaches with its flat *stratus* types lowering; its rain clouds are foglike sheets; when it has passed, it leaves *stratus* wisps, *stratus* fog, or other flat cloud types.

Generally speaking, you can often look overhead and identify the air mass you are in, or the kind of frontal storm that is approaching, by just analyzing the clouds. The drawing showing the typical clouds of the two fronts is worth studying: at first it looks very complicated, but you will find it simple and informative.

These, then, are the two major fronts and their peculiarities. Next, we shall look into examples of each and further compare the kinds of storms that they make.

The Cold-front Storm

A COLD front can be ninety degrees in the shade. A warm front can be ten degrees below zero. They are either "cold" or "warm" only by comparison with the atmosphere they move in upon. Of course, the terms "colder front" and "warmer front" would be more exact.

The effects of the cold-front storm are usually more exciting that any other weather change. It is the summer storm that "clears things off"; it is the winter storm that brings a "nip in the air." It is the liveliest actor in the theater of the sky and it always puts on a good show.

These quick storms that leave colder air in their wake occur all year round, but people seem most impressed by the midsummer examples of cold fronts. On warm, sultry afternoons, after the cumulus clouds have been building up all day long and "there is thunder in the air," you can usually look toward the westerly horizons and see dark mountainous clouds looming there. It doesn't take long for them to come your way; cool air travels fast to wherever it is needed the most. Such air masses constantly sweep the summer skies like careless brooms, leaving cool, dry freshness behind them.

Whereas warm air masses move slowly, cold air masses move fast. They move so much faster than warm air masses that when both travel in the same direction, the cold front often overtakes the warm, climbing up its back and making a higher level (occluded) frontal storm.

You might say the average speed of a brisk cold-front storm is twenty-five to thirty miles an hour. The faster it moves, the more it piles up the frontal clouds, and, as shown in the opposite drawings, concentrated thunderhead masses will often form a few miles apart all along the front.

Whereas the gentle slope of a warm front covers many miles, the steep slant of a cold front is as narrow as it is lively. Fliers are instructed to keep flying straight ahead once they have entered a cold front, and not to turn back. If you continue on, you'll soon break out into the clear again. This narrow band of boisterous cloudform is best described by pictures. The following pages comprise a small gallery of cold-front clouds, some as might be seen from the air.

The stillness before a cold-front storm is a dramatic thing. Hurricanes build up slowly; warm-front clouds increase continually; but the cold-front storm seems to ask for quiet

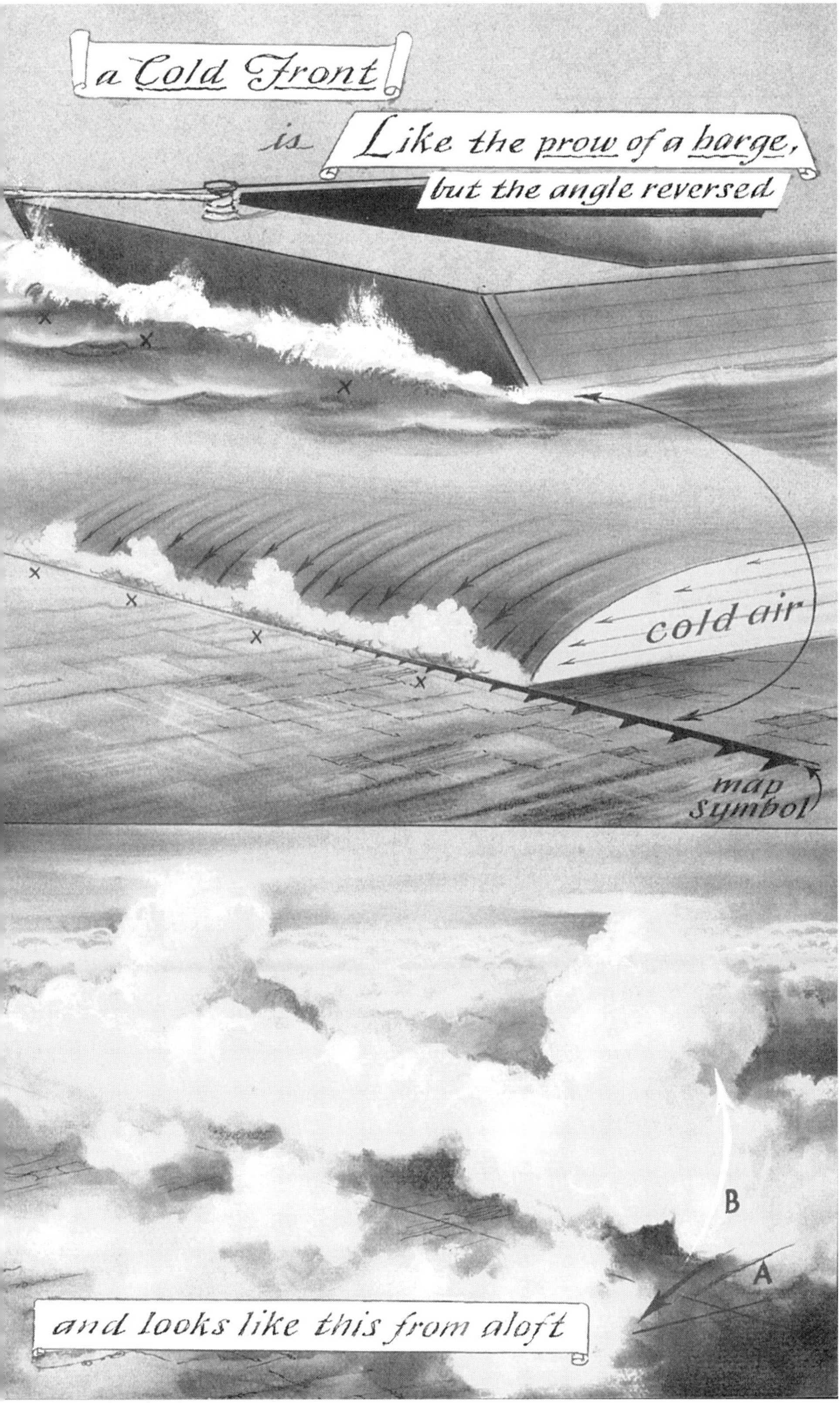

a Cold Front

is *Like the prow of a barge, but the angle reversed*

cold air

map symbol

and looks like this from aloft

B

A

Anatomy of a fast Cold Front storm

Thunderhead

WARM AIR

Scud (cloud fragments)

downdrafts

DIRECTION OF STORM

PURGE OF COOL AIR

Weather map symbol
for Cold Front line

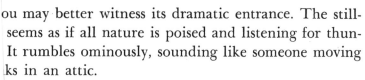

ou may better witness its dramatic entrance. The still-
seems as if all nature is poised and listening for thun-
It rumbles ominously, sounding like someone moving
ks in an attic.

hunder carries for at least ten miles; a fast cold-front
storm moves about thirty miles an hour; so when you first
heard the thunder, the storm should have been ten miles
or twenty minutes away.

The picture shows such a storm moving from the right.
The first turbulent cumulus clouds are already boiling, almost overhead. Peeking from the heights above them, you
can see wisps from the distant thunderhead's anvil top,
streaming out ahead of it. The storm is about five miles
away, and within minutes the quiet will be broken by a stir
of air. The fresh smell of rain will arrive with a shift in the
wind and the stillness before the storm will be over.

squalls build up
flat water areas

In THE cross section through a cold front, you probably noticed a shelf of dark cloud running a little ahead of the storm. That was a line-squall cloud. Sometimes it runs way out ahead, from fifty to a hundred miles in front of the invading cold air. In either case, it is always a dark, straight wall of cloudform in turmoil, pulling overhead like an extended awning before the rain starts. The drawing on the opposite page is a typical fast-moving thunderstorm with its squall cloud ready to flow overhead.

The line-squall cloud is an ideal storm subject for the weather student's observation because it is so low and so active. It usually starts at about 1,600 feet and extends upward another 5,000 feet. It is low enough for you to see the boiling action of mammato cloudform; you can actually hear the howl and whine of the storm above and watch it all at some length before the rain or hail obscures the scene and drives you indoors. Indeed, the awesome blackness of the line squall is one of nature's most impressive sights.

The drawings on the next two pages show the mechanics of two types of line squalls. First is the advance squall (*A*) which rides well ahead, like a "barge wave"; then the roll cloud directly in front of a thunderhead (1 and 2 below). In drawing 1 you see a cold air mass proceeding normally and pushing up cumulus thunderhead formation; next, in drawing 2 the same air mass is slowed by ground friction, causing the nose of the front to "stumble" and topple over into a roll of squall cloud.

The LINE SQUALL
often *rides* *ahead* of a Cold Front
like a *barge-wave*

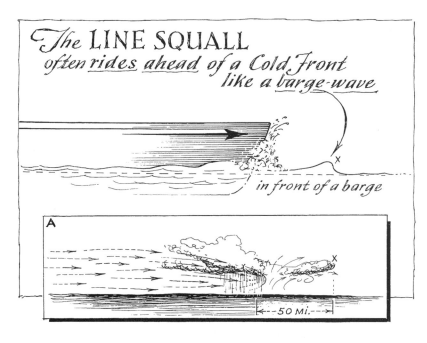

in front of a barge

When the Squall cloud travels WITH (close to) the front, it happens this way

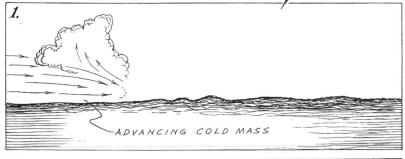

1.

—ADVANCING COLD MASS

2.

LINE SQUALL

Surface speed is checked by ground friction

AND TUMBLES OVER AS AN ADVANCE WAVE

← *diagrams on opposite page.*

"Cold Front Storm reaching its top altitude

The summer thunderstorm with its advance squall is so interesting that you might like to see a diagram showing its mechanism. In the next drawing you see its sequence and the reason why the threatening clouds often boil overhead without letting loose an immediate torrent of rain. In fact, it often passes by and the sky becomes lighter, giving the impression of the storm's having gone. But no—that was just the advance "awning": these squall clouds are only about a mile or less in thickness and they produce little or no rain. Right behind, beneath the cumulo-nimbus, the rain is coming, but hard.

It will be fun to watch for such a storm. Refer to the diagram, and see how closely the actual storm anatomy corresponds to this diagram.

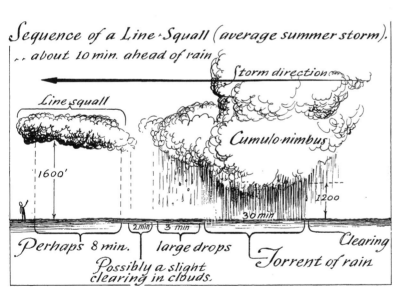

Sequence of a Line-Squall (average summer storm).

The Warm-front Storm

THIS is the storm that gives you most warning. It is usually big and slow: the slant flows forward overhead, so that a complete parade of drab stratus clouds passes by and tells of the rain to come.

The above diagram divides the warm-front sequence into four separate sections, each of which we can then analyze and discuss. Notice that each sequence brings the clouds lower to the ground; finally in drawing 4 the frontal line meets the earth level in a deluge of precipitation.

The slant of the front (which we already remarked as being slight) is necessarily exaggerated here. To portray such a long, slight slant in correct proportions, the page of this book would have to be forty feet wide! For example of its size and gentle slope, a warm-front storm flowing from the west would be pushing cirrus clouds (sequence 1) over New York while the other end of its slant (sequence 4) would be pouring rain in Chicago; such a sloping front would extend a thousand miles.

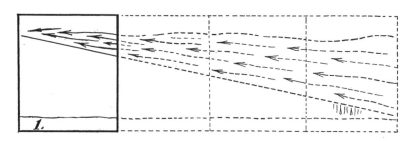

MARE'S-TAIL" clouds have long been known as signs of storm. The most well-known mariner's folklore has it that:

> "Mackerel skies and mare's tails,
> Means batten down and shorten sails."

Actually, neither mackerel sky (cirro-cumulus clouds) nor mare's-tails (cirrus wisps) have any definite forecast significance. The finest summer day will often have its sky graced with big cirrus plumes. Mare's-tails are just a sign of brisk wind to come, but not necessarily a sign of storm. Only when cirrus ceases to be confined to patches and starts to shroud the whole sky is a storm on its way.

May we first explain cirrus for those who are unfamiliar with the word. Over five miles up, all cloudform is cirrus in type, being composed of *ice*. That sounds heavy, but don't worry about being hit on the head with blocks of ice; cirrus clouds are formed of very tiny ice particles.

Stretched across the expanse of the upper sky there are always fields of these ice clouds so fine and transparent that the sun and moon shine right through them. Only when the sun is low (as in rising or setting) and shines against cirrus indirectly, blazing the sky with scarlet, do these sheets of cloudform become entirely visible.

Warm air is light and it rises quickly. When a large stormy mass of it is on the move, the first warning is where its "nose" has risen to icy heights, shrouding the sky with cobwebby sheets.

Cirrostratus
Webby sky...rain to come.

Cirrus mare's tails
Scattered wisps...wind to come.

The drawing shows the first sign of a warm-front storm (upper drawing) as compared with fair-weather cirrus mare's-tails in the lower drawing.

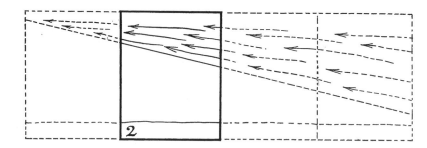

A COBWEBBY cirrus sky usually changes from its substratosphere heights (about 35,000 feet) to cirro-stratus clouds at about 30,000 feet within an hour or two. This lowering of altitude is unnoticeable to the eye, but cirro-stratus which is a more solid sheet than cirrus, is thick enough to cause sun or moon halos, which identify it.

Sky halos are exactly like the halos you see through "angel's-hair" on a Christmas tree around each light bulb. The same effect would also occur if you shone a flashlight through a cake of ice. Sky halos are actually light shining through ice.

Winter halos are more frequent and less of a storm warning, but summertime halos are more often the sun or moon shining through the ice particles of a warm-front storm which should reach you in about twelve hours. The drawing below explains the mechanics of a sky halo.

about 30,000 ft.

CIRROSTRATUS

ice cloud produces Halo

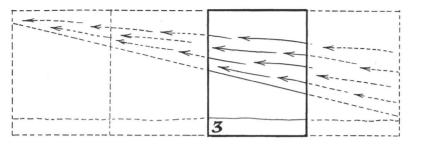

Sᴜɴ and moon halos disappear when the warm-front sequence lowers and thickens into alto-stratus cloud-form. When the sun or moon shines through this cloud ceiling, it appears like a ground glass disk; it also loses its shape and you cannot tell whether it is round or square. Other than this identifying quality, it is difficult to define the alto-stratus cloud except that it is a shapeless, dull and drab ceiling, like a lowering fog. Early folklore has it that "When sun and moon become a blur, within six hours rains will stir."

Altostratus

Ground glass effect, shapeless sun or moon.

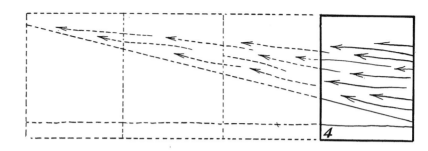

Fourth Sequence of a Warm Front

By NOW most of the long slanting front has glided past overhead; its lower end is about to reach the level of earth. The clouds are low enough to loosen rainfall, and the slow drizzle begins. The average warm-front rain is usually long-lasting but less torrential than that of the cold-front storm. The old-timers say that it lasts as long "as it took to come"—which would mean if it took twelve hours for the four cloud sequences to pass overhead, the rain would also last twelve hours.

The last stage of the warm-front storm does not lend itself to pictorial illustration because there is just dull, drab rainfall and a sullen sky. The movement and lift of air is usually too slow and gradual for violent storm action, so the warm front produces fewer thunderstorms. But warm-front thunderstorms do evolve, particularly along coastal areas where the frontal angle is apt to become steeper, and where cold masses linger to do battle with the invading warm air.

The warm-front thunderstorm is the persistent, long-lasting, less-violent, foggy, sultry storm with lightning that seems to be far away. Actually, the lightning *is* far away—it seldom strikes ground because it is happening so far above you. The drawing on the following page shows the heights of such a storm, the warm air flowing up a "mountain slope" of colder air. Notice how the thunderheads are formed far above the front and how lightning occurs between clouds rather than from the clouds to the ground.

Cross-section through a Warm Front Thunderstorm

Warm Air-mass

air in wake of storm is hazy, dull.

Lifted-air Storms

ONE last trip back to our "oar and water" demonstration. We come to the third way of making a storm disturbance—that of lifting. The drawing on the opposite page shows three sample ways that a mass of storm-ingredient atmosphere can be lifted fast enough and high enough to produce a full-fledged thunderstorm.

Just like the steam that issues from a boiling teapot, clouds and storms are caused by a cooling of air. Hence storm air must first be warm. The way it cools is usually by lifting itself away from the warmth of the earth. We might say that thunderstorms depend upon: (1) warm air, (2) rising.

Wherever atmosphere rises rapidly, thunderstorms prevail. Wet air rises faster, so warm wet air is natural storm stuff. You'd think that wet air would be heavy and less liable to rise; but no, wet air is lighter!

The *thermal thunderstorm* (*A*), which is the most common type, is just an overgrown cumulus formation. Instead of cumulus thermals building up and dying down normally, they are here fed by surrounding thermals until the circulation creates a monster of a thunderstorm cell.

Thermal thunderstorms usually form in the afternoon, travel with the prevailing wind, and die out in the late afternoon. The wetter the air, however, the later into the evening such a storm will continue.

The *air-mass storm* (*B*) is due to unstable air combined with an overrunning of cold air aloft. The drawing shows how a high jet-stream of cold air seems to suck the thunderhead currents upward, aiding to build up its circulation. Such storms are milder, with less cloud-to-ground lightning strikes unless they occur in mountainous areas.

The *orographical storm* (*C*) is fed by air lifted along land slopes as wind. Such a thunderstorm may remain almost stationary, hovering over an elevated place, forming and reforming as long as the wet, warm wind continues to blow uphill, lifting the air.

Storms made by Lifting...

...3 common ways that atmosphere rises

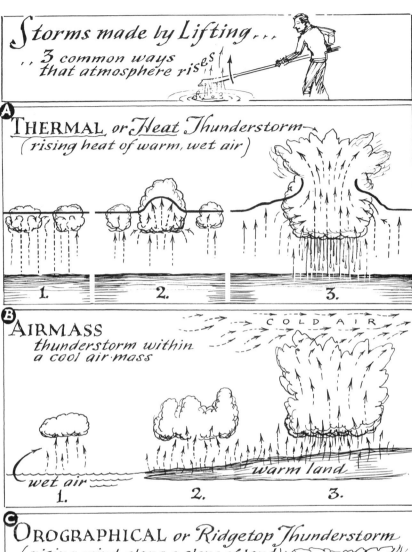

A. THERMAL, or Heat Thunderstorm
(rising heat of warm, wet air)

1.   2.   3.

B. AIRMASS
thunderstorm within a cool air·mass

COLD AIR

wet air

warm land

1.   2.   3.

C. OROGRAPHICAL or Ridgetop Thunderstorm
(rising wind along a slope of land)

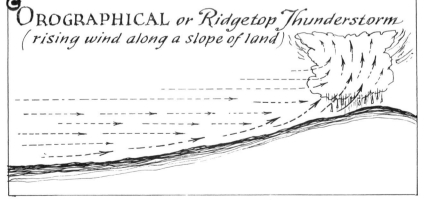

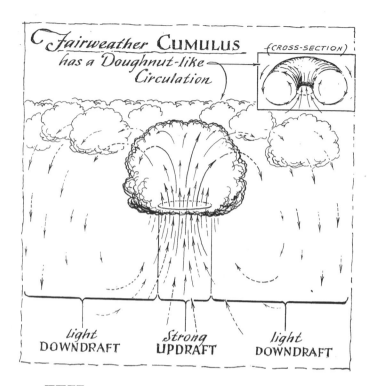

Fairweather CUMULUS has a Doughnut-like Circulation (CROSS-SECTION)

Light
DOWNDRAFT

Strong
UPDRAFT

Light
DOWNDRAFT

Thunder-storm Clouds

W E HAVE referred to the cumulo-nimbus, or thunderhead, as "an overgrown fair-weather cumulus cloud." Many weather books do the same thing, and it is often confusing to the student. How can the fairest of fair-weather clouds (the cumulus) become the meanest of mean clouds (the cumulo-nimbus) just by growing big? You'd think by getting bigger, it would become even more "fair"! Although the distinction between the two clouds has seldom been dwelt upon, we shall try to do so here. The difference is not only in size, it seems, but also in a change of mechanics.

The fluffy, white fair-weather cumulus cloud has a dough-nut sort of circulation with air funneling upward in a center chimney and then falling gently all around it. This aerial machine forms long enough to create a cumulus-cloud formation, and then ceases with the disappearance of the cloud.

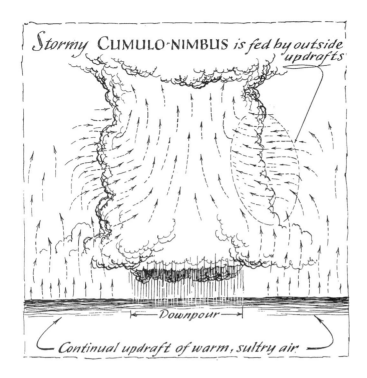

Stormy CUMULO-NIMBUS *is fed by outside updrafts*

—— Downpour ——

Continual updraft of warm, sultry air

It may be said that a cumulus cloud is the visible top of an invisible cell of thermal circulation.

The cumulo-nimbus, however, does not die away so soon, for it is fed by the abundance of surrounding thermals of rising warm, wet air. Its mechanics involves the cloud being fed from the sides, as the drawing shows.

The growth of a thunderhead from the cumulus beginning is a dramatic thing to witness, although its action is so slow that it can only be seen by means of time-lapse motion pictures. The following pages give an idea of the way an air-mass or heat thunderstorm grows from a cumulus to a cumulo-nimbus mountain of cloudform.

The first sign of cumulus clouds becoming possible storm material is when they take on a chaotic appearance, when they lose their oval shape, when they lose their nice mathematical placement in the over-all pattern of cumulus-cloud fields, and when they grow castlelike turrets.

The turreted cloud called cumulus-castellatus or its higher brother, the alto-cumulus-castellatus, cannot be detected from beneath, but when they are at a distance or seen from an airplane, their anatomy is unmistakable. As their shape suggests, they are formed by a rapid upward rush of air. They are children of the unstable warm sultry afternoon and forerunners of the summer shower. As folklore puts it:

> "When clouds appear like hills and towers,
> The earth's refreshed by frequent showers."

Altocumulus Castellatus
The Castle Cloud

oval crown

Castellatus

flat base

Continued fair

Storm ahead

The Beginning of a Thunderhead

FLYING over a field of cumulus or alto-cumulus on a hot summer afternoon, you can often spot a dozen or more towered clouds that might become thunderheads within a few hours.

Here we see such a cloud, rising out of its prairie of cumulus clouds and reaching the critical altitude of about 7,000 feet. The dotted line indicates how much it will probably grow. The increasing mass of cloud at this point might be termed "cumulus congestus."

Cumulo-nimbus being born from a field of Cumulus clouds

7000'

3500'

Birth of a Thunder-head

Here we see the side-fed circulation that departs from that of a cumulus cloud and starts the thunderhead on its way. A close-flying sailplane could be sucked into a cloud like this and gain altitude as if it were in an elevator.

This cloud is just about reaching the freezing level sufficient to produce large raindrops; most of the rain falling beneath it up until now will evaporate before it reaches the earth.

Cumulo-nimbus building up

On Its
Way to
Becoming
a Storm

HERE the thunderhead has pushed far above the cumulus fields below it and rain is falling steadily from beneath it. A larger thunderhead is disintegrating at the near left and fragments of its anvil top are drifting into the scene.

Airplanes do not enter clouds this size without expecting a severe buffeting. The air-mass thunderhead is usually so confined that it can be easily flown around. The plane shown here is flying between two thunderheads, avoiding turbulent air as much as it can.

young Thunderhead still growing,
in grand-dad's shadow

**The Great
Storm
Cloud**

No SIGHT is more majestic or awesome than that of a full-grown cumulo-nimbus. Higher than any earth mountain, it towers into the substratosphere and dwarfs even the horizon.

The thunderhead shown here has reached its peak. A scarf of cirruslike foam cloud is developing around its anvil top, and from here on, the cloud will slowly begin to disintegrate. The internal circulation at this point is changing from ascending currents to descending ones. But the turbulence inside is just as forbidding. At a distance the thunderhead seems motionless and as solid as marble, but as you fly close to it you see a foaming, boiling texture, alive with the glow and flash of electrical charge from within.

Grand·dad himself...
a full grown Cumulo·nimbus

**The
Umbrella
of Hail**

THE anvil top of a giant thunderhead is more a shower of hail and snow than the harmless foggy veil it appears to be. Airplanes that have flown beneath it have been bombarded by giant hailstones and buffeted by descending currents.

The thunderhead often scratches the floor of the stratosphere, which is about five miles high at the poles and about ten miles high at the equator. (The atmosphere swings outward by centrifugal force, making it thicker at the equator than at the poles.) In temperate zones, thunderheads average about 35,000 feet high, but pilots flying in equatorial areas have reported flying at 50,000 feet while the anvil tops of thunderheads loomed "ten thousand feet above."

Dissipating Anvil-top

Thunder-storms and Radar

BELOW, you see the dial of a weather radarscope and the theory of its working. Modern airplanes can fly through thunderstorms without damage, and can even be hit by lightning (without harm) but the ride through turbulent air is not pleasant. Now, by turning radar to the heaviest concentration of rain and snow, those spots which are recognized as the most turbulent can be spotted and flown around.

The drawing on the opposite page shows the weather radar of a plane searching out the storm cells of roughest air so that the flight course may be changed to fly around or between them. The process is like a bus driver watching for bumps in the road and avoiding them. Cells *A, B,* and *C* show up on the radarscope in their exact positions, while the circles show their distance (in miles) away from the plane.

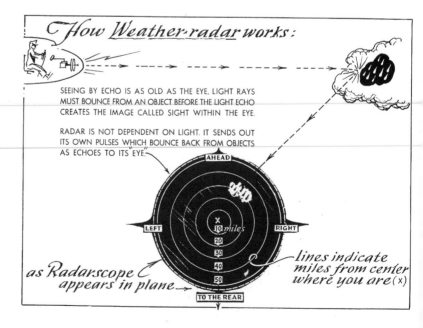

How *Weather-radar works:*

SEEING BY ECHO IS AS OLD AS THE EYE. LIGHT RAYS MUST BOUNCE FROM AN OBJECT BEFORE THE LIGHT ECHO CREATES THE IMAGE CALLED SIGHT WITHIN THE EYE.

RADAR IS NOT DEPENDENT ON LIGHT. IT SENDS OUT ITS OWN PULSES WHICH BOUNCE BACK FROM OBJECTS AS ECHOES TO ITS "EYE."

AHEAD

LEFT

RIGHT

X 10 *miles*

20

30

40

50

TO THE REAR

lines indicate miles from center where you are (x)

as Radarscope C appears in plane

How weather-radar spots storm-centers

detour

These cells of roughest air

contain snow and big raindrops which show on the scope like this

in the plane

RADARSCOPE.

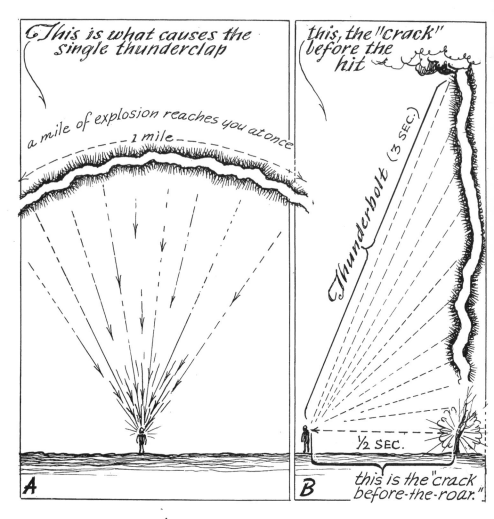

This is what causes the single thunderclap

a mile of explosion reaches you at once

— 1 mile —

this, the "crack" before the hit

Thunderbolt (3 SEC.)

½ SEC.

this is the "crack before-the-roar."

A

B

Thunder   A BOLT of lightning is, as far as our conception goes, instantaneous; but sound is slow and the noise from the mile or two of explosion takes various lengths of time to reach our ears. The result is that roaring rumble of noise called thunder.

Above, you see two types of thunder strikes: the "thunderclap" and the "thunderbolt." The noise from the thunderclap (*A*) reaches your ears all at one time because the explosion was arched over your head, all of it the same distance away. The thunderbolt (B) is a bolt which has aimed toward you, striking nearby: first you hear the noise between you and the place struck, which will sound like a very loud fraction-of-a-second "snap" or "click," followed by the noise from the bolt itself, possibly three seconds long.

The distance of lightning can be told easily by counting the seconds between the flash and the first sound of thunder. Dividing that by five will give you the distance in miles. You can multiply the seconds by 1,000 feet and get the same effect, but you will get your distance in thousands of feet. You don't need a watch; just say, "This second is one, this second is two, this second is three," and you will get a fairly accurate timing. Five seconds (5 x 1000) will be 5,000 feet, or nearly a mile.

If you wanted to count the length of the lightning flash itself, you would have to count the length of the thunder's rumble. Did I hear you say you'd be measuring an echo? Most people think that, too. The rumble of thunder contains very little echo, because you will hear the same thing on the prairie or at sea, where there are no hills to cause an echo. Lightning is a long and irregular river of light, so its noise is naturally a long and irregular river of sound. By counting the seconds of the rumble you will not necessarily be measuring the whole length of the flash, but you will be measuring the lengths away from you the river of light had extended. When the flash runs sideways to you, as shown in the next drawing as "BANG!" you get a whole length of the noise at once and the effect is called a "thunderclap." Many people think that a big thunderclap means a strike to earth, but by looking at the drawing, you see that a thunderclap can be just that part of an overhead bolt which runs parallel to the ground.

Thunder cannot be heard for more than twenty-five miles in most favorable air currents, but usually only ten miles away. This accounts for what we call "heat lightning," or distant lightning without thunder. Of course, this is re-

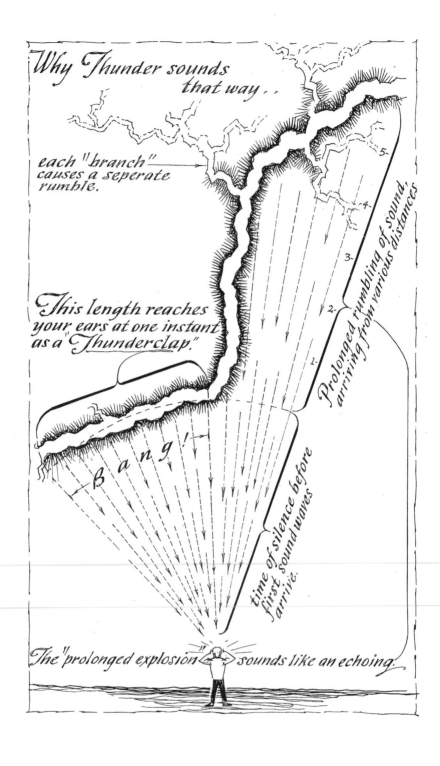

Sheet lightning
(internal discharge)

flected lightning, usually at night, too far away for the thunder to reach you. Why it should be called heat lightning is perhaps because more distant storms might occur during warm nights, although actually noise does carry less far in warmed air. *Sheet lightning* is another term for the distant reflection of lightning below the horizon, but its most common use is to describe the reflected light of lightning which is entirely within a cloud.

Although you only see the lightning that leaps out of a thunderhead, there is much discharge occurring within its depths which you cannot see. However, if you could fly close to a thunderhead, you would see an almost constant flickering from within, giving the scary effect of a great fire going on inside. Such a flickering of sheet lightning is the visible effect of the disturbing atmospherics which make your radio or TV ineffective when a storm is nearby.

Your best bet against lightning is in a
Metal frame building or one with Rods

2.
In a "Faraday Cage"

3. *In a ravine, in a ditch, against a cliff, in thick forests, or lying flat on the ground*

LIGHTNING is as freakish about striking close by without injuring a person as it is about singling out its target. You are in possible danger during a thunderstorm if you are in an open automobile or in a partly metal airplane; yet you are safest if the automobile or airplane is entirely metal-enclosed. Halfway in a car, on a bicycle, tractor, or other metal conveyance is not good sense. A wire fence can carry a fatal charge for a mile.

Lightning

If you are skittish about electrical phenomena, you might avoid screen doors, fireplaces, open windows, the center of the room, or being in contact with any metal objects while in the house. You are not actually attracting lightning, but a close hit might give you a reactionary jolt. Taking a bath or being in contact with plumbing during lightning is inviting disaster. People who are struck while swimming or in contact with wetness are not always killed by burns; rather, their lungs and diaphragm are paralyzed with shock. The same artificial respiration used for drownings can often revive them.

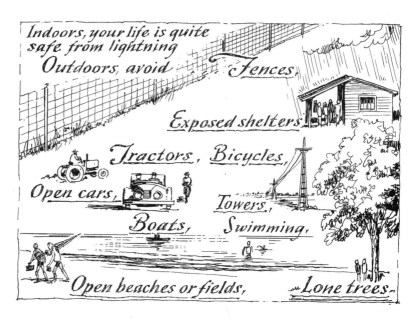

Indoors, your life is quite safe from lightning
Outdoors, avoid — Fences,
Exposed shelters,
Tractors, Bicycles,
Open cars,
Towers,
Boats, Swimming,
Open beaches or fields,
Lone trees.

The best-known place of safety during a thunderstorm is inside an automobile: the usual supposition is that rubber tires offer the protection. Of course, water is a conductor of electricity and in the rain, wet rubber tires would give very little safety from lightning current. The true protection comes from what is known as the Faraday Cage principle. It was Michael Faraday who found that an enclosed metal container protects its insides from any outside electrical disturbance.

In 1836 Faraday built himself a twelve-foot box and covered it with sheets of tin foil. He entered and lived within it, making his experiments. Using all the electricity-producing gadgets known to his time, he found that the outside of his cubicle could be alive with sparks while the inside remained normal.

A lightning bolt may strike an all-metal automobile or airplane without harming its occupants and very often does so. Very few long-time pilots are not familiar with lightning bolts; it is odd that up there where lightning starts, and where hits are most frequent, there is more safety than in an unprotected place down on the earth.

Folklore has it that groups of animals through their heat and quantity of charged fur attract lightning. This, according to science, is highly improbable, although when a barn is hit, all the animals within are usually a direct target for the bolt. And it is perfectly true that large groups of cows, horses, or sheep have often been struck, although tall trees, which should have made better targets, were surrounding them. This theory will probably always be confined to folklore, but the rise and fall of electrical potential on the ground and in earthly objects as a storm passes overhead is nevertheless well known. As it approaches, the ground gradient is first increased positively, then reversed for an instant beneath the first squall line, increased again directly beneath the thunderhead and finally predominantly negative underneath the rest of the storm cloud. The drawing shows the ground gradient being changed from positive to negative, according to the electrical charge that hovers over-

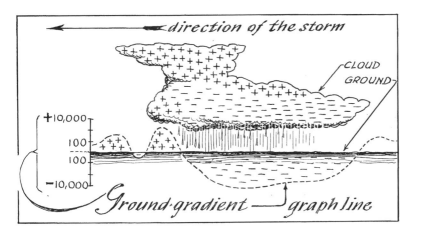

direction of the storm

CLOUD
GROUND

+10,000
100
100
−10,000

Ground-gradient ——— graph line

head. Nearly all the strokes to the ground, incidentally, involve the negative charge toward the rear of the clouds.

It is strange how little progress man has made toward some aspects of weather. Although we have mastered the use of electricity, and are capable of making man-made lightning, our protection against it has not changed much since the time of George Washington: the lightning rod we use today is quite the same. Benjamin Franklin made extensive experiments with lightning, even making research into ground and cloud gradients. One of his electrometers or measuring devices for electric current was his electric chimes such as the one drawn here. Mention of it is made in a letter to Collinson of London: "September, 1752. After the kite experiment, I erected an iron rod to draw the lightning

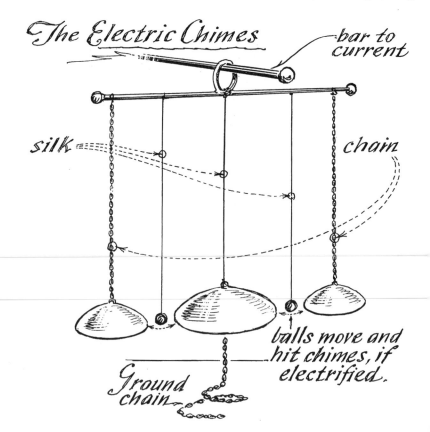

The Electric Chimes

bar to current

silk

chain

balls move and hit chimes, if electrified.

Ground chain

down into my house, in order to make some experiments on it, with two bells to give notice when the rod should be electrified; a contrivance obvious to every electrician. I found the bells rang sometimes when there was no lightning or thunder, but only a dark cloud over the rod. . . . That sometimes . . . when they had not rung before they would, after a flash, suddenly begin to ring." The electrical chimes that Franklin referred to were more important than his kite and key as a means of researching electricity. The process was equally dangerous. A few months after the above letter was written, Professor George William Richman set up a rod to lead lightning into his study. He did it so well that an electric charge passed from the instrument to his head, killing him instantly.

Ball lightning is seen on rare occasions, but optic phenomena have often been what we mistook for ball lightning. Just as when you glance at a sunlit windshield of an approaching automobile, and the flash stays with you for several seconds after, lightning will often take the form of a lingering ball of light in your aftervision. The effect of "dancing" occurs as you shift your line of sight from here to there, a few seconds after the flash. Ball lightning is a strange and terrifying thing to behold, and only about as many have actually witnessed it as have been struck or nearly struck by ordinary lightning.

True ball lightning has never been explained. Accounts have been recorded of lightning bolts breaking into two or more "balls of fire" after striking an object, then rolling or flying slowly up or down chimneys, or dancing about the room before exploding. Although there is no proof to date of its being so, ball lightning is thought to be a magnetically held-together concentration of corona discharge, such as exists in "St. Elmo's light."

St. Elmo's fire, or St. Elmo's light, is a distant cousin of lightning, but in aviation the relationship becomes very close. Imagine yourself in an airplane, flying through thundery weather, at high altitude. Eerie lights brighten the fog-like cirrus; the crackling on your radio almost eliminates any other signals. It is evident that the air is alive with electrical disturbances. Finally there is no other signal but the loud rasping of outside interference; your ship is collecting an overload of current. You switch the useless radio off, and knowing your weather, watch for the overload to build up and cause a glow from the tips of the wings. Sure enough, there it starts; first a blue-white tinge, then a yellowish glow. It's nothing to worry about—it's just St. Elmo's fire. They will glow and disappear; or if you keep collecting the over-

CONCENTRATION OF POSITIVE CURRENT CAUSED BY UPDRAFT

PLANE WITH EQUAL LOAD

PLANE COLLECTS OVERCHARGE

load, you will discharge it into the next cloud you pass that can use it.

But wait—you *are* collecting more; now the "flames" dance up and down the wing, and the propeller tips are churning through a solid sheet of orange light! Time to put your red glasses on, and wait for the discharge. It will be actual lightning, you know, discharging from your plane into a cloud, and without red glasses you would be temporarily blinded. BANG! There it goes! Your overload went out like a shot into a passing cloud. The noise was exactly like thunder because it *was* thunder. You have witnessed St.

GLOWS WITH *St. Elmo's Fire*

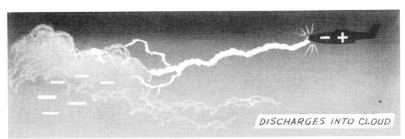

DISCHARGES INTO CLOUD

Elmo's fire at its peak, and lightning at its lowest form. It was exciting, but there was no damage, and nothing to have worried about at any time.

Lightning is a product of vertical winds. The horizontal wind of a hurricane seldom produces thunderstorm material, yet even a small-area ascent of warm summer air can separate the electrical charge of that part of the sky, so that the unequal charges begin to bolt back and forth in their effort to equalize.

There are usually enough uneven charges within the thundercloud, so that lightning bolts will discharge aloft or between clouds; but a small percentage of the bolts discharge into the earth. The process is shown in the set of diagrams where the overcharged cloud has sent down a "pilot streamer" in search of a suitable "emptying place." If you were about to be struck, you might possibly feel your pilot streamer reaching toward the cloud's pilot streamer the second before the final contact and discharge. There are records of linemen who while working on high poles during a storm suddenly felt a bristling sensation and found their hair standing on end. By quickly leaving the pole and falling flat on the ground, they avoided the lightning bolt that followed.

THE next three pages will show the steps taken within an instantaneous flash of lightning, too fast for the human eye to see, but proven by high-speed cameras. It is an analysis of the bolt shown above.

The lightning bolt is similar in action (and remarkably the same in looks) to a river. It is like an overflow of water, pouring from many tributary streams, all emptying into the main river which flows directly to the sea. A high-altitude camera picture of a river, and a picture of a lightning bolt are exactly alike in anatomy.

The following pictures show one action of a strike to the ground, but a single flash may contain as many as forty or fifty similar repeat actions.

The Mechanics of a Lightning Bolt

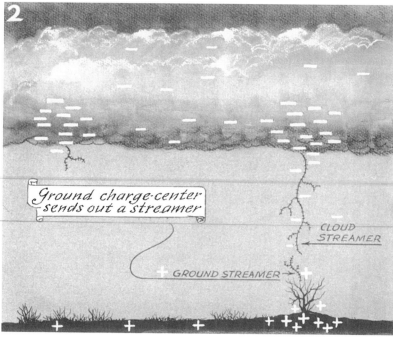

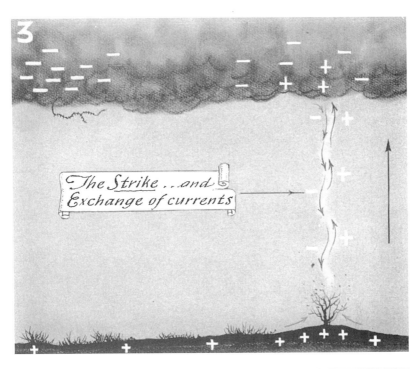

The Strike...and Exchange of currents

One charge·center is completely discharged,

but

Streamers develop between the two Charge·centers.

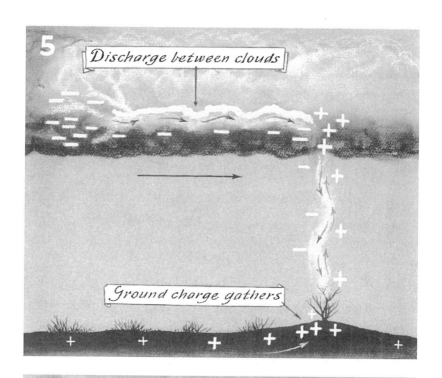

5 Discharge between clouds

Ground charge gathers

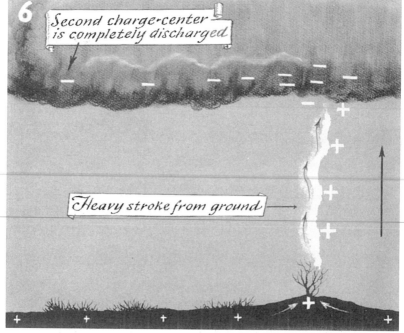

6 Second charge-center is completely discharged

Heavy stroke from ground

At the instant of lightning strike, the first charge actually goes from the ground to the cloud, rather than the cloud's striking the earth. Thereafter, there are several exchanges of current going first up the bolt then down until the electrical potentials between cloud and ground have equalized. In the case of a high object, the initial streamer most often starts from the tall object. Out of fifty-two strikes on the Empire State Building in New York, the building had "struck the cloud" fifty times! The graph

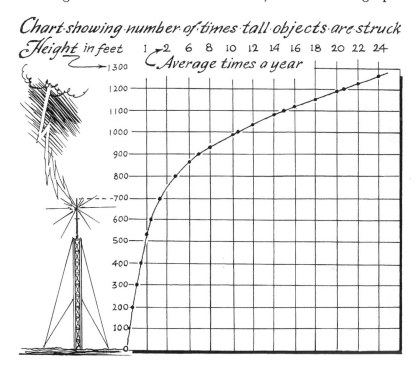

Chart showing number of times tall objects are struck

Height in feet   1 2 6 8 10 12 14 16 18 20 22 24

Average times a year

showing the relationship between height of an object and its susceptibility to being hit was copied from *Lightning Phenomena* by Westinghouse's Mr. G. D. McCann. It further states that transmission lines ranging from seventy to a hundred and ten feet high are struck once a year per mile of line.

On the following pages is a schematic diagram showing six

Six kinds of Lightning numbered according to frequency of occurrence

1. Within clouds, from front to back of storm
2. Within clouds, from upper clouds to lower clouds
3. "Glow discharge" into surrounding atmosphere

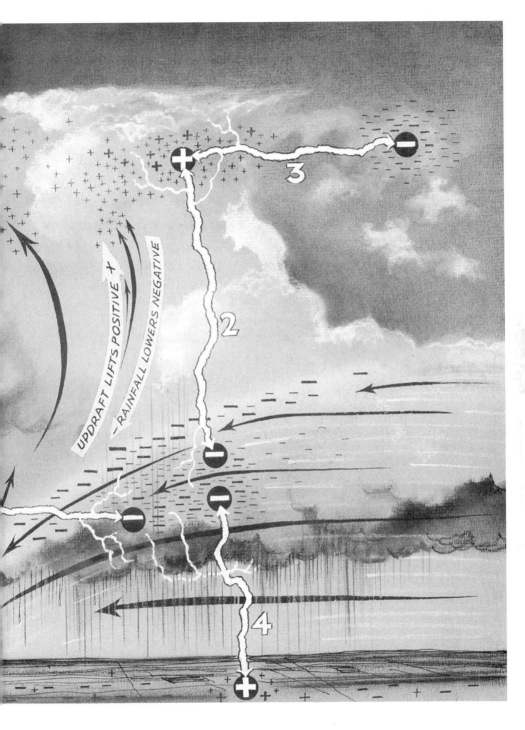

4. From low rain clouds to earth (greatest danger)
5. From squall cloud to earth (infrequent)
6. From upper cloud to earth (rare)

types of lightning, numbered according to their frequency of occurrence. The general pattern of the thundercloud is that ascending winds carry positive current aloft while descending rains carry negative current downward. The exact reasons why the electrical potentials should separate to begin with are still unproven. Friction, the action of wind against raindrops, the falling of rain from great heights, and many other reasons have scientific foundation; perhaps it is all of them put together that causes the final material for lightning. When you drive on dry pavements with chains on your automobile, and the static collects enough to make the car radio useless; when you walk across a rug and collect a charge that sparks to anything you touch; when you comb your hair and cracking sparks dance along your comb—all these phenomena are kin to the lightning bolt.

UNDERSTANDING the mechanics of storms is only **Conclusion** the beginning of the study of weather, but it is a sound basis. As you might have noticed, there has not been mention, in this book, of rain or hail or snow. Although they are the ingredients of storm, the creation of such precipitation is not as well understood as one might think. Vincent Schaefer, who is one of America's leading meteorologists and recognized as the expert on rain-making, states, "We do not yet know much about the processes which cause rain or snow to form in clouds." Even the simplest phenomena of storm precipitation is understood only in theory, with new interpretations arising from the scientific world all the time.

But without delving into the mathematics of meteorology, the layman can find enjoyment watching the panorama of sky and being aware of the unseen currents and air masses that actuate the machinery of weather. All around you, there are miniature storms happening, in your teapot or within the air circulation of your room, with warm air rising and cold air descending. Why the fireplace "backfires" and the hall is chilly will be no mystery to the one who is weatherwise and aware of the movements of air. When the weather map is vague and the weather report uncertain, the changing sky will contain information for a more accurate local weather prediction. A little weather wisdom adds a lot to the joy of living.

A CATALOG OF SELECTED
DOVER BOOKS
IN ALL FIELDS OF INTEREST

A CATALOG OF SELECTED DOVER
BOOKS IN ALL FIELDS OF INTEREST

CONCERNING THE SPIRITUAL IN ART, Wassily Kandinsky. Pioneering work by father of abstract art. Thoughts on color theory, nature of art. Analysis of earlier masters. 12 illustrations. 80pp. of text. 5⅜ x 8½.　　　　0-486-23411-8

CELTIC ART: The Methods of Construction, George Bain. Simple geometric techniques for making Celtic interlacements, spirals, Kells-type initials, animals, humans, etc. Over 500 illustrations. 160pp. 9 x 12. (Available in U.S. only.)　　　　0-486-22923-8

AN ATLAS OF ANATOMY FOR ARTISTS, Fritz Schider. Most thorough reference work on art anatomy in the world. Hundreds of illustrations, including selections from works by Vesalius, Leonardo, Goya, Ingres, Michelangelo, others. 593 illustrations. 192pp. 7⅛ x 10¼.　　　　0-486-20241-0

CELTIC HAND STROKE-BY-STROKE (Irish Half-Uncial from "The Book of Kells"): An Arthur Baker Calligraphy Manual, Arthur Baker. Complete guide to creating each letter of the alphabet in distinctive Celtic manner. Covers hand position, strokes, pens, inks, paper, more. Illustrated. 48pp. 8¼ x 11.　　　　0-486-24336-2

EASY ORIGAMI, John Montroll. Charming collection of 32 projects (hat, cup, pelican, piano, swan, many more) specially designed for the novice origami hobbyist. Clearly illustrated easy-to-follow instructions insure that even beginning papercrafters will achieve successful results. 48pp. 8¼ x 11.　　　　0-486-27298-2

BLOOMINGDALE'S ILLUSTRATED 1886 CATALOG: Fashions, Dry Goods and Housewares, Bloomingdale Brothers. Famed merchants' extremely rare catalog depicting about 1,700 products: clothing, housewares, firearms, dry goods, jewelry, more. Invaluable for dating, identifying vintage items. Also, copyright-free graphics for artists, designers. Co-published with Henry Ford Museum & Greenfield Village. 160pp. 8¼ x 11.　　　　0-486-25780-0

THE ART OF WORLDLY WISDOM, Baltasar Gracian. "Think with the few and speak with the many," "Friends are a second existence," and "Be able to forget" are among this 1637 volume's 300 pithy maxims. A perfect source of mental and spiritual refreshment, it can be opened at random and appreciated either in brief or at length. 128pp. 5⅜ x 8½.　　　　0-486-44034-6

JOHNSON'S DICTIONARY: A Modern Selection, Samuel Johnson (E. L. McAdam and George Milne, eds.). This modern version reduces the original 1755 edition's 2,300 pages of definitions and literary examples to a more manageable length, retaining the verbal pleasure and historical curiosity of the original. 480pp. 5³⁄₁₆ x 8¼.　　　　0-486-44089-3

ADVENTURES OF HUCKLEBERRY FINN, Mark Twain, Illustrated by E. W. Kemble. A work of eternal richness and complexity, a source of ongoing critical debate, and a literary landmark, Twain's 1885 masterpiece about a barefoot boy's journey of self-discovery has enthralled readers around the world. This handsome clothbound reproduction of the first edition features all 174 of the original black-and-white illustrations. 368pp. 5⅜ x 8½.　　　　0-486-44322-1

STICKLEY CRAFTSMAN FURNITURE CATALOGS, Gustav Stickley and L. & J. G. Stickley. Beautiful, functional furniture in two authentic catalogs from 1910. 594 illustrations, including 277 photos, show settles, rockers, armchairs, reclining chairs, bookcases, desks, tables. 183pp. 6½ x 9¼. 0-486-23838-5

AMERICAN LOCOMOTIVES IN HISTORIC PHOTOGRAPHS: 1858 to 1949, Ron Ziel (ed.). A rare collection of 126 meticulously detailed official photographs, called "builder portraits," of American locomotives that majestically chronicle the rise of steam locomotive power in America. Introduction. Detailed captions. xi+ 129pp. 9 x 12. 0-486-27393-8

AMERICA'S LIGHTHOUSES: An Illustrated History, Francis Ross Holland, Jr. Delightfully written, profusely illustrated fact-filled survey of over 200 American lighthouses since 1716. History, anecdotes, technological advances, more. 240pp. 8 x 10¾. 0-486-25576-X

TOWARDS A NEW ARCHITECTURE, Le Corbusier. Pioneering manifesto by founder of "International School." Technical and aesthetic theories, views of industry, economics, relation of form to function, "mass-production split" and much more. Profusely illustrated. 320pp. 6⅛ x 9¼. (Available in U.S. only.) 0-486-25023-7

HOW THE OTHER HALF LIVES, Jacob Riis. Famous journalistic record, exposing poverty and degradation of New York slums around 1900, by major social reformer. 100 striking and influential photographs. 233pp. 10 x 7⅞. 0-486-22012-5

FRUIT KEY AND TWIG KEY TO TREES AND SHRUBS, William M. Harlow. One of the handiest and most widely used identification aids. Fruit key covers 120 deciduous and evergreen species; twig key 160 deciduous species. Easily used. Over 300 photographs. 126pp. 5⅜ x 8½. 0-486-20511-8

COMMON BIRD SONGS, Dr. Donald J. Borror. Songs of 60 most common U.S. birds: robins, sparrows, cardinals, bluejays, finches, more—arranged in order of increasing complexity. Up to 9 variations of songs of each species.
Cassette and manual 0-486-99911-4

ORCHIDS AS HOUSE PLANTS, Rebecca Tyson Northen. Grow cattleyas and many other kinds of orchids—in a window, in a case, or under artificial light. 63 illustrations. 148pp. 5⅜ x 8½. 0-486-23261-1

MONSTER MAZES, Dave Phillips. Masterful mazes at four levels of difficulty. Avoid deadly perils and evil creatures to find magical treasures. Solutions for all 32 exciting illustrated puzzles. 48pp. 8¼ x 11. 0-486-26005-4

MOZART'S DON GIOVANNI (DOVER OPERA LIBRETTO SERIES), Wolfgang Amadeus Mozart. Introduced and translated by Ellen H. Bleiler. Standard Italian libretto, with complete English translation. Convenient and thoroughly portable—an ideal companion for reading along with a recording or the performance itself. Introduction. List of characters. Plot summary. 121pp. 5¼ x 8½. 0-486-24944-1

FRANK LLOYD WRIGHT'S DANA HOUSE, Donald Hoffmann. Pictorial essay of residential masterpiece with over 160 interior and exterior photos, plans, elevations, sketches and studies. 128pp. 9¼ x 10¾. 0-486-29120-0

THE CLARINET AND CLARINET PLAYING, David Pino. Lively, comprehensive work features suggestions about technique, musicianship, and musical interpretation, as well as guidelines for teaching, making your own reeds, and preparing for public performance. Includes an intriguing look at clarinet history. "A godsend," *The Clarinet*, Journal of the International Clarinet Society. Appendixes. 7 illus. 320pp. 5⅜ x 8½. 0-486-40270-3

HOLLYWOOD GLAMOR PORTRAITS, John Kobal (ed.). 145 photos from 1926-49. Harlow, Gable, Bogart, Bacall; 94 stars in all. Full background on photographers, technical aspects. 160pp. 8⅜ x 11¼. 0-486-23352-9

THE RAVEN AND OTHER FAVORITE POEMS, Edgar Allan Poe. Over 40 of the author's most memorable poems: "The Bells," "Ulalume," "Israfel," "To Helen," "The Conqueror Worm," "Eldorado," "Annabel Lee," many more. Alphabetic lists of titles and first lines. 64pp. 5$^{3}/_{16}$ x 8¼. 0-486-26685-0

PERSONAL MEMOIRS OF U. S. GRANT, Ulysses Simpson Grant. Intelligent, deeply moving firsthand account of Civil War campaigns, considered by many the finest military memoirs ever written. Includes letters, historic photographs, maps and more. 528pp. 6⅛ x 9¼. 0-486-28587-1

ANCIENT EGYPTIAN MATERIALS AND INDUSTRIES, A. Lucas and J. Harris. Fascinating, comprehensive, thoroughly documented text describes this ancient civilization's vast resources and the processes that incorporated them in daily life, including the use of animal products, building materials, cosmetics, perfumes and incense, fibers, glazed ware, glass and its manufacture, materials used in the mummification process, and much more. 544pp. 6$^{1}/_{8}$ x 9$^{1}/_{4}$. (Available in U.S. only.) 0-486-40446-3

RUSSIAN STORIES/RUSSKIE RASSKAZY: A Dual-Language Book, edited by Gleb Struve. Twelve tales by such masters as Chekhov, Tolstoy, Dostoevsky, Pushkin, others. Excellent word-for-word English translations on facing pages, plus teaching and study aids, Russian/English vocabulary, biographical/critical introductions, more. 416pp. 5⅜ x 8½. 0-486-26244-8

PHILADELPHIA THEN AND NOW: 60 Sites Photographed in the Past and Present, Kenneth Finkel and Susan Oyama. Rare photographs of City Hall, Logan Square, Independence Hall, Betsy Ross House, other landmarks juxtaposed with contemporary views. Captures changing face of historic city. Introduction. Captions. 128pp. 8¼ x 11. 0-486-25790-8

NORTH AMERICAN INDIAN LIFE: Customs and Traditions of 23 Tribes, Elsie Clews Parsons (ed.). 27 fictionalized essays by noted anthropologists examine religion, customs, government, additional facets of life among the Winnebago, Crow, Zuni, Eskimo, other tribes. 480pp. 6⅛ x 9¼. 0-486-27377-6

TECHNICAL MANUAL AND DICTIONARY OF CLASSICAL BALLET, Gail Grant. Defines, explains, comments on steps, movements, poses and concepts. 15-page pictorial section. Basic book for student, viewer. 127pp. 5⅜ x 8½. 0-486-21843-0

THE MALE AND FEMALE FIGURE IN MOTION: 60 Classic Photographic Sequences, Eadweard Muybridge. 60 true-action photographs of men and women walking, running, climbing, bending, turning, etc., reproduced from rare 19th-century masterpiece. vi + 121pp. 9 x 12. 0-486-24745-7

ANIMALS: 1,419 Copyright-Free Illustrations of Mammals, Birds, Fish, Insects, etc., Jim Harter (ed.). Clear wood engravings present, in extremely lifelike poses, over 1,000 species of animals. One of the most extensive pictorial sourcebooks of its kind. Captions. Index. 284pp. 9 x 12. 0-486-23766-4

1001 QUESTIONS ANSWERED ABOUT THE SEASHORE, N. J. Berrill and Jacquelyn Berrill. Queries answered about dolphins, sea snails, sponges, starfish, fishes, shore birds, many others. Covers appearance, breeding, growth, feeding, much more. 305pp. 5¼ x 8¼. 0-486-23366-9

ATTRACTING BIRDS TO YOUR YARD, William J. Weber. Easy-to-follow guide offers advice on how to attract the greatest diversity of birds: birdhouses, feeders, water and waterers, much more. 96pp. 5³⁄₁₆ x 8¼. 0-486-28927-3

MEDICINAL AND OTHER USES OF NORTH AMERICAN PLANTS: A Historical Survey with Special Reference to the Eastern Indian Tribes, Charlotte Erichsen-Brown. Chronological historical citations document 500 years of usage of plants, trees, shrubs native to eastern Canada, northeastern U.S. Also complete identifying information. 343 illustrations. 544pp. 6½ x 9¼. 0-486-25951-X

STORYBOOK MAZES, Dave Phillips. 23 stories and mazes on two-page spreads: Wizard of Oz, Treasure Island, Robin Hood, etc. Solutions. 64pp. 8¼ x 11. 0-486-23628-5

AMERICAN NEGRO SONGS: 230 Folk Songs and Spirituals, Religious and Secular, John W. Work. This authoritative study traces the African influences of songs sung and played by black Americans at work, in church, and as entertainment. The author discusses the lyric significance of such songs as "Swing Low, Sweet Chariot," "John Henry," and others and offers the words and music for 230 songs. Bibliography. Index of Song Titles. 272pp. 6½ x 9¼. 0-486-40271-1

MOVIE-STAR PORTRAITS OF THE FORTIES, John Kobal (ed.). 163 glamor, studio photos of 106 stars of the 1940s: Rita Hayworth, Ava Gardner, Marlon Brando, Clark Gable, many more. 176pp. 8⅜ x 11¼. 0-486-23546-7

YEKL and THE IMPORTED BRIDEGROOM AND OTHER STORIES OF YIDDISH NEW YORK, Abraham Cahan. Film Hester Street based on *Yekl* (1896). Novel, other stories among first about Jewish immigrants on N.Y.'s East Side. 240pp. 5⅜ x 8½. 0-486-22427-9

SELECTED POEMS, Walt Whitman. Generous sampling from *Leaves of Grass*. Twenty-four poems include "I Hear America Singing," "Song of the Open Road," "I Sing the Body Electric," "When Lilacs Last in the Dooryard Bloom'd," "O Captain! My Captain!"–all reprinted from an authoritative edition. Lists of titles and first lines. 128pp. 5³⁄₁₆ x 8¼. 0-486-26878-0

SONGS OF EXPERIENCE: Facsimile Reproduction with 26 Plates in Full Color, William Blake. 26 full-color plates from a rare 1826 edition. Includes "The Tyger," "London," "Holy Thursday," and other poems. Printed text of poems. 48pp. 5¼ x 7. 0-486-24636-1

THE BEST TALES OF HOFFMANN, E. T. A. Hoffmann. 10 of Hoffmann's most important stories: "Nutcracker and the King of Mice," "The Golden Flowerpot," etc. 458pp. 5⅜ x 8½. 0-486-21793-0

THE BOOK OF TEA, Kakuzo Okakura. Minor classic of the Orient: entertaining, charming explanation, interpretation of traditional Japanese culture in terms of tea ceremony. 94pp. 5⅜ x 8½. 0-486-20070-1

CATALOG OF DOVER BOOKS

FRENCH STORIES/CONTES FRANÇAIS: A Dual-Language Book, Wallace Fowlie. Ten stories by French masters, Voltaire to Camus: "Micromegas" by Voltaire; "The Atheist's Mass" by Balzac; "Minuet" by de Maupassant; "The Guest" by Camus, six more. Excellent English translations on facing pages. Also French-English vocabulary list, exercises, more. 352pp. 5⅜ x 8½. 0-486-26443-2

CHICAGO AT THE TURN OF THE CENTURY IN PHOTOGRAPHS: 122 Historic Views from the Collections of the Chicago Historical Society, Larry A. Viskochil. Rare large-format prints offer detailed views of City Hall, State Street, the Loop, Hull House, Union Station, many other landmarks, circa 1904-1913. Introduction. Captions. Maps. 144pp. 9⅜ x 12¼. 0-486-24656-6

OLD BROOKLYN IN EARLY PHOTOGRAPHS, 1865-1929, William Lee Younger. Luna Park, Gravesend race track, construction of Grand Army Plaza, moving of Hotel Brighton, etc. 157 previously unpublished photographs. 165pp. 8⅞ x 11¾. 0-486-23587-4

THE MYTHS OF THE NORTH AMERICAN INDIANS, Lewis Spence. Rich anthology of the myths and legends of the Algonquins, Iroquois, Pawnees and Sioux, prefaced by an extensive historical and ethnological commentary. 36 illustrations. 480pp. 5⅜ x 8½. 0-486-25967-6

AN ENCYCLOPEDIA OF BATTLES: Accounts of Over 1,560 Battles from 1479 B.C. to the Present, David Eggenberger. Essential details of every major battle in recorded history from the first battle of Megiddo in 1479 B.C. to Grenada in 1984. List of Battle Maps. New Appendix covering the years 1967-1984. Index. 99 illustrations. 544pp. 6½ x 9¼. 0-486-24913-1

SAILING ALONE AROUND THE WORLD, Captain Joshua Slocum. First man to sail around the world, alone, in small boat. One of great feats of seamanship told in delightful manner. 67 illustrations. 294pp. 5⅜ x 8½. 0-486-20326-3

ANARCHISM AND OTHER ESSAYS, Emma Goldman. Powerful, penetrating, prophetic essays on direct action, role of minorities, prison reform, puritan hypocrisy, violence, etc. 271pp. 5⅜ x 8½. 0-486-22484-8

MYTHS OF THE HINDUS AND BUDDHISTS, Ananda K. Coomaraswamy and Sister Nivedita. Great stories of the epics; deeds of Krishna, Shiva, taken from puranas, Vedas, folk tales; etc. 32 illustrations. 400pp. 5⅜ x 8½. 0-486-21759-0

MY BONDAGE AND MY FREEDOM, Frederick Douglass. Born a slave, Douglass became outspoken force in antislavery movement. The best of Douglass' autobiographies. Graphic description of slave life. 464pp. 5⅜ x 8½. 0-486-22457-0

FOLLOWING THE EQUATOR: A Journey Around the World, Mark Twain. Fascinating humorous account of 1897 voyage to Hawaii, Australia, India, New Zealand, etc. Ironic, bemused reports on peoples, customs, climate, flora and fauna, politics, much more. 197 illustrations. 720pp. 5⅜ x 8½. 0-486-26113-1

THE PEOPLE CALLED SHAKERS, Edward D. Andrews. Definitive study of Shakers: origins, beliefs, practices, dances, social organization, furniture and crafts, etc. 33 illustrations. 351pp. 5⅜ x 8½. 0-486-21081-2

THE MYTHS OF GREECE AND ROME, H. A. Guerber. A classic of mythology, generously illustrated, long prized for its simple, graphic, accurate retelling of the principal myths of Greece and Rome, and for its commentary on their origins and significance. With 64 illustrations by Michelangelo, Raphael, Titian, Rubens, Canova, Bernini and others. 480pp. 5⅜ x 8½. 0-486-27584-1

PSYCHOLOGY OF MUSIC, Carl E. Seashore. Classic work discusses music as a medium from psychological viewpoint. Clear treatment of physical acoustics, auditory apparatus, sound perception, development of musical skills, nature of musical feeling, host of other topics. 88 figures. 408pp. 5⅜ x 8½. 0-486-21851-1

LIFE IN ANCIENT EGYPT, Adolf Erman. Fullest, most thorough, detailed older account with much not in more recent books, domestic life, religion, magic, medicine, commerce, much more. Many illustrations reproduce tomb paintings, carvings, hieroglyphs, etc. 597pp. 5⅜ x 8½. 0-486-22632-8

SUNDIALS, Their Theory and Construction, Albert Waugh. Far and away the best, most thorough coverage of ideas, mathematics concerned, types, construction, adjusting anywhere. Simple, nontechnical treatment allows even children to build several of these dials. Over 100 illustrations. 230pp. 5⅜ x 8½. 0-486-22947-5

THEORETICAL HYDRODYNAMICS, L. M. Milne-Thomson. Classic exposition of the mathematical theory of fluid motion, applicable to both hydrodynamics and aerodynamics. Over 600 exercises. 768pp. 6⅛ x 9¼. 0-486-68970-0

OLD-TIME VIGNETTES IN FULL COLOR, Carol Belanger Grafton (ed.). Over 390 charming, often sentimental illustrations, selected from archives of Victorian graphics–pretty women posing, children playing, food, flowers, kittens and puppies, smiling cherubs, birds and butterflies, much more. All copyright-free. 48pp. 9¼ x 12¼. 0-486-27269-9

PERSPECTIVE FOR ARTISTS, Rex Vicat Cole. Depth, perspective of sky and sea, shadows, much more, not usually covered. 391 diagrams, 81 reproductions of drawings and paintings. 279pp. 5⅜ x 8½. 0-486-22487-2

DRAWING THE LIVING FIGURE, Joseph Sheppard. Innovative approach to artistic anatomy focuses on specifics of surface anatomy, rather than muscles and bones. Over 170 drawings of live models in front, back and side views, and in widely varying poses. Accompanying diagrams. 177 illustrations. Introduction. Index. 144pp. 8⅜ x11¼. 0-486-26723-7

GOTHIC AND OLD ENGLISH ALPHABETS: 100 Complete Fonts, Dan X. Solo. Add power, elegance to posters, signs, other graphics with 100 stunning copyright-free alphabets: Blackstone, Dolbey, Germania, 97 more–including many lower-case, numerals, punctuation marks. 104pp. 8⅛ x 11. 0-486-24695-7

THE BOOK OF WOOD CARVING, Charles Marshall Sayers. Finest book for beginners discusses fundamentals and offers 34 designs. "Absolutely first rate . . . well thought out and well executed."–E. J. Tangerman. 118pp. 7¾ x 10⅝. 0-486-23654-4

ILLUSTRATED CATALOG OF CIVIL WAR MILITARY GOODS: Union Army Weapons, Insignia, Uniform Accessories, and Other Equipment, Schuyler, Hartley, and Graham. Rare, profusely illustrated 1846 catalog includes Union Army uniform and dress regulations, arms and ammunition, coats, insignia, flags, swords, rifles, etc. 226 illustrations. 160pp. 9 x 12. 0-486-24939-5

WOMEN'S FASHIONS OF THE EARLY 1900s: An Unabridged Republication of "New York Fashions, 1909," National Cloak & Suit Co. Rare catalog of mail-order fashions documents women's and children's clothing styles shortly after the turn of the century. Captions offer full descriptions, prices. Invaluable resource for fashion, costume historians. Approximately 725 illustrations. 128pp. 8⅜ x 11¼. 0-486-27276-1

LIGHT AND SHADE: A Classic Approach to Three-Dimensional Drawing, Mrs. Mary P. Merrifield. Handy reference clearly demonstrates principles of light and shade by revealing effects of common daylight, sunshine, and candle or artificial light on geometrical solids. 13 plates. 64pp. 5⅜ x 8½. 0-486-44143-1

ASTROLOGY AND ASTRONOMY: A Pictorial Archive of Signs and Symbols, Ernst and Johanna Lehner. Treasure trove of stories, lore, and myth, accompanied by more than 300 rare illustrations of planets, the Milky Way, signs of the zodiac, comets, meteors, and other astronomical phenomena. 192pp. 8⅜ x 11. 0-486-43981-X

JEWELRY MAKING: Techniques for Metal, Tim McCreight. Easy-to-follow instructions and carefully executed illustrations describe tools and techniques, use of gems and enamels, wire inlay, casting, and other topics. 72 line illustrations and diagrams. 176pp. 8¼ x 10⅞. 0-486-44043-5

MAKING BIRDHOUSES: Easy and Advanced Projects, Gladstone Califf. Easy-to-follow instructions include diagrams for everything from a one-room house for bluebirds to a forty-two-room structure for purple martins. 56 plates; 4 figures. 80pp. 8¾ x 6⅜. 0-486-44183-0

LITTLE BOOK OF LOG CABINS: How to Build and Furnish Them, William S. Wicks. Handy how-to manual, with instructions and illustrations for building cabins in the Adirondack style, fireplaces, stairways, furniture, beamed ceilings, and more. 102 line drawings. 96pp. 8¾ x 6⅜. 0-486-44259-4

THE SEASONS OF AMERICA PAST, Eric Sloane. From "sugaring time" and strawberry picking to Indian summer and fall harvest, a whole year's activities described in charming prose and enhanced with 79 of the author's own illustrations. 160pp. 8¼ x 11. 0-486-44220-9

THE METROPOLIS OF TOMORROW, Hugh Ferriss. Generous, prophetic vision of the metropolis of the future, as perceived in 1929. Powerful illustrations of towering structures, wide avenues, and rooftop parks–all features in many of today's modern cities. 59 illustrations. 144pp. 8¼ x 11. 0-486-43727-2

THE PATH TO ROME, Hilaire Belloc. This 1902 memoir abounds in lively vignettes from a vanished time, recounting a pilgrimage on foot across the Alps and Apennines in order to "see all Europe which the Christian Faith has saved." 77 of the author's original line drawings complement his sparkling prose. 272pp. 5⅜ x 8½. 0-486-44001-X

THE HISTORY OF RASSELAS: Prince of Abissinia, Samuel Johnson. Distinguished English writer attacks eighteenth-century optimism and man's unrealistic estimates of what life has to offer. 112pp. 5⅜ x 8½. 0-486-44094-X

A VOYAGE TO ARCTURUS, David Lindsay. A brilliant flight of pure fancy, where wild creatures crowd the fantastic landscape and demented torturers dominate victims with their bizarre mental powers. 272pp. 5⅜ x 8½. 0-486-44198-9

Paperbound unless otherwise indicated. Available at your book dealer, online at **www.doverpublications.com**, or by writing to Dept. GI, Dover Publications, Inc., 31 East 2nd Street, Mineola, NY 11501. For current price information or for free catalogs (please indicate field of interest), write to Dover Publications or log on to **www.doverpublications.com** and see every Dover book in print. Dover publishes more than 500 books each year on science, elementary and advanced mathematics, biology, music, art, literary history, social sciences, and other areas.